"Exasperated by eggs? Challenged by cheese? Your life is about to change, thanks to *The Complete Guide to Vegan Food Substitutions*! Fun and informative, the book discusses the science of cooking and baking and shows how easy it is to embrace a cruelty-free lifestyle. Filled with more than 200 tried-and-true recipes, this book is essential for any vegan or aspiring vegan's bookshelf."

—TAMASIN NOYES, author of *American Vegan Kitchen*

"Fun and informative, *The Complete Guide to Vegan Food Substitutions* is a creative feast for the eyes, fuel for imagination, and most importantly, a true palate pleaser. Expect to be entertained and satisfied by this one-of-a-kind cookbook."

—ALISA FLEMING, author of *Go Dairy Free: The Guide and Cookbook for Milk Allergies, Lactose Intolerance, and Casein-Free Living* (www.GoDairyFree.com)

"It is not just right, it is beautiful; it is inspired and it makes a perfect gift for those relatives who always ask what on earth you eat!"

—JEFFREY MASSON, PH.D., bestselling author of *When Elephants Weep, The Face on Your Plate: The Truth About Food,* and *The Dog Who Couldn't Stop Loving*

"*The Complete Guide to Vegan Food Substitutions* is a goldmine for anyone desiring to understand the practical keys to healthier, kinder, more planet-friendly eating. I have found it to be a treasure trove of substitution gems to confidently veganize traditional recipes with no sacrifice of taste or texture, and it's packed with user-friendly recipes and beautiful photos to boot. Not only are there are sections on the savvy substitution of dairy, egg, meat, and animal by-products, but for gluten, soy, sugar, and fat as well. At last we have a practical manual to reclaim nutritional sanity!"

—WILL TUTTLE, PH.D., pianist, composer, former Zen monk, author of the acclaimed bes
and recipient of the C

This book is dedicated to all who are working, in their own big or small way, on making the world a far gentler and more respectful habitat for the animal kingdom.

Text © 2011 Celine Steen and Joni Marie Newman
Photography © 2011 Fair Winds Press

First published in the USA in 2011 by
Fair Winds Press, a member of
Quayside Publishing Group
100 Cummings Center
Suite 406-L
Beverly, MA 01915-6101
www.fairwindspress.com

15 14 13 12 11   3 4 5

ISBN-13: 978-1-59233-441-4
ISBN-10: 1-59233-441-5

Library of Congress Cataloging-in-Publication Data

Steen, Celine.
  The complete guide to vegan food substitutions : Veganize it! Foolproof methods
for transforming any dish into a delicious new vegan favorite / Celine Steen and
Joni Marie Newman.
      p. cm.
  Includes bibliographical references and index.
  ISBN-13: 978-1-59233-441-4
  ISBN-10: 1-59233-441-5
  1. Meat substitutes. 2. Food substitutes. I. Newman, Joni-Marie. II. Title.
  TX838.S74 2010
  641.5'636—dc22

                        2010028369

Cover design: Nancy Ide Bradham, bradhamdesign.com
Book design: carol holtz | holtzdesign.com
Photography: Celine Steen | havecakewilltravel.com

Printed and bound in China

INCLUDES RECIPES!

THE COMPLETE GUIDE TO VEGAN FOOD SUBSTITUTIONS

Veganize It! Foolproof Methods for Transforming Any Dish into a Delicious New Vegan Favorite

Celine Steen and Joni Marie Newman
Authors of 500 Vegan Recipes
Photography by Celine Steen

FAIR WINDS
PRESS
BEVERLY, MASSACHUSETTS

Contents

Introduction

SO HERE YOU ARE, leafing through *The Complete Guide to Vegan Food Substitutions.* Welcome! No matter what your reasons for wanting a change, the fact remains that you're doing something for your health, the good of the animals, as well as the planet.

A guide to vegan food substitutions has never been more needed, considering the number of people who are choosing to make lifestyle changes for ethical or environmental reasons. Equally important are those who have no choice but to completely overhaul their eating habits due to health issues.

With so many ways to replace traditional ingredients with those that are free of animal products and allergens, you will soon wonder why anyone would even bother with the less healthy, not cruelty-free versions in the first place.

We created this book to unveil the mysteries of vegan food substitutions, to make it possible for the everyday home cook to create dairy-free, egg-free, meat-free, and animal-free dishes that are both delicious and kind.

You may have family recipes going back several generations and feel like giving them a healthier, animal-friendly spin. While you could simply go out and try to find a vegan cookbook that contains a somewhat similar recipe, there is enormous and lasting satisfaction that comes from knowing that pretty much any dish can be made vegan using your own hands and your own set of cooking and baking skills. That's the goal of our book: to give you the know-how and confidence to transform any traditional recipe you come across— and have the dish turn out delicious, healthy, and kind every single time.

Throughout this book, we will teach you how to make your own, or shop for simple ingredient swaps that "veganize" your favorite recipes. We will give you recommendations on particular products and brands that have performed well in our kitchens. Then, at the start of each chapter, we will walk you through a traditional recipe and point out precisely what we would do to make it vegan. Before you know it, you'll have that same ability— allowing you to look at any recipe and instantly translate it for the vegan chef.

We're big fans of making our food from scratch. It's usually cheaper and so important to know exactly what goes into our food. As a bonus, we can also take full credit for the outcome! That being said, we find that the quality of some items have yet to be rivaled in the home kitchen, so, for instance, we stick to the store-bought kind of cream cheese and other melt-y cheeses. And in other instances, we're just too lazy to bother making things like tofu and tempeh.

Enough preambles; let's get started on this tasty trip into the magical world of vegan food substitutions! (Unicorns not included.)

Purely Pound Cake (page 192)

A Note on Food Allergies

We cannot stress enough how vital it is that you keep an extremely close eye on the ingredients you use before making a recipe that is labeled as being gluten-, nut-, soy-free, and so on.

While we have done our best to label said recipes, allergens still have a way of sneaking into ingredients that you'd expect to be respectful of the restrictions you have to follow. You are your best advocate. We care about your health, so please use extreme caution.Throughout this book you will find recipes labeled in the following way:

Corn Free: Recipes that do not contain corn.

For the More Experienced Cook: Recipes that take a little more time and are best enjoyed on days when you have more freedom. Or, that could be prepared in several steps, on several different occasions, to cut back on cooking and/or baking time.

Gluten Free: Recipes that do not contain gluten. Double-check ingredient labels. Recipes that contain certain ingredients such as vinegars, grain alcohol, and flavor extracts (such as vanilla) may be labelled as gluten free. Use your best judgment when consuming these products, and search out certified gluten free ingredients when preparing foods for those with gluten sensitivity. For more information regarding gluten free ingredients, visit Celiac.com.

Low Fat: Recipes that contain three grams (the equivalent of 27 fat calories) of fat, or less, per serving.

Nut Free: Recipes that do not contain nuts. Double-check ingredient labels.

Quick and Easy: Recipes that take less than thirty minutes to whip up, provided you have intermediate cooking and/or baking skills.

Raw: Recipes that only contain raw ingredients or ingredients not heated to temperatures above 118°F (48°C).

Soy Free: Recipes that do not contain soy. Double-check ingredient labels.

Wheat Free: Recipes that do not contain wheat. Double-check ingredient labels.

Throughout the book you will also see the following symbols, which highlight the main substitute in each recipe, according to the subject matter of that chapter, so that you can easily detect what works where and how:

Dairy Sub

Cheese Sub

Egg Sub

Meat Sub

Fish Sub

By-product Sub

Gluten Sub

Soy Sub

Sugar Sub

Fat Sub

Ingredients

As you flip through the pages of this book, you may come across ingredients you are not be familiar with. That's okay; it's all part of the process. Most of the ingredients used in this book can be found in any supermarket—and stores make it easy on us vegans by grouping most of the vegan cheeses, yogurts, faux-meats, mayos, and other items together, usually in the produce department alongside the tofu. The less common ingredients may require a trip to your local health-food store or co-op, and if you're unable to find a particular ingredient in your area, a quick Internet search should give you many options.

Below is a list of ingredients that may require extra explanation and a brief description of how we use them throughout this book. More in-depth descriptions of certain ingredients appear in each of the chapters.

Beans: For convenience's sake, we like to use canned beans, rinsing and draining them thoroughly to get rid of unnecessary extra sodium. If you prefer cooking your own, more power to you! Keep in mind that one 15-ounce (425 g) can of beans generally equals approximately 1$\frac{2}{3}$ cups (294 g) cooked beans, or $\frac{2}{3}$ cup (120 g) dry beans.

Chia Seeds: Small mild-flavored seeds, usually dark gray or black in color, that become gelatinous when combined with liquid, making them an excellent thickening agent in puddings, dressings, and even crackers.

Flours: Scooping vs. lightly spooning? We're of the lightly spooning and scraping school. This means that we use a spoon to transfer flours into the measuring cup, so as not to overpack the stuff. It can make a difference in how recipes turn out, so it's a good thing to keep in mind.

As for the kind of flour we prefer, we like to get as much nutrition out of our food as we possibly can, no matter how sinful or healthy the recipe. That's why we generally use whole wheat pastry flour for our baked goods. If whole wheat pastry flour is hard to find, you can simply combine equal amounts of unbleached all-purpose and whole wheat flours to get the same results. Of course, you can go with all-purpose flour only, if you want baked goods that are truest to form.

Liquid smoke: A flavoring that's stocked near the marinades in most markets. Actually made by condensing smoke into liquid form. A little goes a long way in giving a smoky flavor to many foods.

Maca powder: A dried root that is somewhat of a superfood, since it is packed with vitamins (B), minerals (including calcium), and amino acids, and is said to increase stamina, reduce fatigue, and even enhance libido. It is rather expensive, but a little goes a long way.

Nondairy milks: We like ours to be unsweetened so that it lends itself to pretty much any use, but we aren't too picky on which kind: soy, almond, rice, hemp . . . They're all tasty but you will have to do a taste-testing for yourself, as both brands and personal preference might play a role here. When it comes to making buttermilk for baking by combining nondairy milk and vinegar or lemon juice, we find that soy is the best option. Please have a look at chapter 1 (page 16) for more on nondairy milks.

Nut butters: We love them all, as long as they're made of nothing but nuts or at most have a little bit of added salt. Please have a look at chapter 1 (page 17) for more on nut butters.

Nutritional yeast: The magical, nonactive kind of yeast most vegans adore. Its nutty and cheesy flavor is a bit of an acquired taste, so give yourself time to get used to it, and you might find yourself hopelessly addicted to it like many others before you. Look for the vegetarian-support formula (this will be noted on the label), as it is enriched in vitamin B_{12}.

Oils: We generally use neutral-flavored vegetable oils (such as canola, vegetable, peanut, etc.) in baking, unless otherwise mentioned.

Canola oil has a bit of a bad reputation due to GMO issues, so you can opt for peanut or vegetable oil, but the choice is up to you and depends on what you can afford or what your health conscience tells you. When it comes to cooking, we like to use extra-virgin olive oil on salads, and peanut oil for dishes that involve a long frying time as it has a high-smoke point.

There are two camps when it comes to coconut oil: Some people think it's highly unhealthy, while others swear by it and think it can cure pretty much any ailment—and save your marriage, too. We're all for moderation and actually like the stuff, so you will find a few recipes in this book that list it as an ingredient: Again, do what your conscience and wallet tell you, and substitute away! It is, after all, the purpose of this book to show you that recipes do not have to be rigid, and that you can develop your own guidelines for what sounds and tastes right to you.

Salt and **pepper:** We like to respect your habits when it comes to salt and pepper, so the measurements you will find in our recipes are meant as a guide. We usually add "to taste" so that you can follow your needs and preferences.

We prefer using sea salt, as it retains a minuscule amount of minerals. And we like to use a small amount of black salt in recipes that replicate eggs, to lend them a delicate sulfurous flavor.

Seaweed, such as **hijiki, dulse,** and **nori:** Edible seaweeds add a fishy flavor to foods without using fish.

Soy sauce: Can be replaced with tamari or Bragg Liquid Aminos. If you're watchful of your sodium intake, purchase the reduced-sodium kind. The liquid aminos only contain a small amount of natural sodium and happen to be gluten free. The flavor and cost, however, may take a little getting used to.

Sriracha or **"Rooster Sauce":** An addictive hot sauce made from chili peppers, garlic, vinegar, and salt ground together to form a smooth paste. Check for ingredients, as some brands contain fish sauce.

Sweeteners: We have used a wide selection of sweeteners in this book to show you their different uses and qualities. We most commonly use evaporated cane juice (in place of granulated white sugar), brown sugar, Sucanat, and raw sugar. As for the liquid sweeteners, we're big fans of agave nectar, pure maple syrup, and brown rice syrup. We have also dabbled in calorie-free all-natural sweeteners such as stevia and xylitol. Please have a look at chapter 8 (page 226) to get more information about our most commonly used types of sweetening agents.

Tamarind paste: We love to use this sour condiment, but the type we find at the international food store isn't made of 100 percent tamarind and therefore a bit milder than pure tamarind paste. Adjust amounts according to what you can find and to your liking.

Tofu and **tempeh:** Silken, soft, medium, firm, extra firm, even super firm! We use silken for desserts and sauces, and prefer firm or even firmer to fry or bake. Keep in mind it's best to press your tofu if you're going to fry it, as it makes for a chewier, meatier texture. Here's an easy method: drain any liquid from the block of tofu and then sandwich it between either two folded kitchen towels or several folded layers of paper towels and place a heavy object, such as a book or frying pan, on top to press out excess moisture. Allow to sit for about an hour to press. Please have a look at chapter 4 (page 109) for more information about tofu and tempeh.

Vital wheat gluten: A perfect source of protein to make delicious seitan (the glorious wheat meat), or used in breads to improve texture, gluten is the natural protein portion removed from whole wheat. It can be found in most grocery stores these days or ordered online. It is important to know that vital wheat gluten flour is completely different from high gluten flour. The two are not interchangeable and will not perform similarly in recipes.

Xanthan gum: Don't let the name scare you: This natural carbohydrate that is produced from corn or cabbage does wonders for gluten-free baking and cooking by adding volume to the goods. Available at most health-food stores or online, one bag should last you a very long time.

Please have a look at chapter 6 (page 173) for more information about xanthan gum.

LET THE COWS COME HOME!

FOOLPROOF SUBSTITUTIONS FOR DAIRY

Chapter 1

From Frozen Delights to Dreamy Sauces:
HOW TO SUBSTITUTE FOR DAIRY

DID YOU KNOW that the human race is the only species on earth that consumes another mammal's milk for sustenance? How's that for spooky?

Consider the Facts

Milk's primary function is to provide high levels of nutrition, including fat, protein, carbohydrate, and calcium, to a newborn baby through breastfeeding before the baby is capable of consuming other foods.

In nature, milk is meant to be consumed only by the offspring of the mother who is producing it. Milk contains high levels of casein, which is a protein known to have opiate-like effects. The purpose of this is to cause an addictive response in the infant so it will crave the mother's milk and continue to feed and nourish.

Lactose is the sugar found in milk. In infants, the intestinal villi produce lactase, an enzyme secreted specifically to break down lactose. As the infant grows, the amount of naturally produced lactase decreases, making it more and more uncomfortable to digest. Is it any wonder that so many people are lactose intolerant? It simply isn't natural or healthy to consume another mammal's milk.

Guidelines for Substituting for Dairy

Unlike many other foods, substituting for dairy is pretty straightforward. For the most part, if a recipe calls for 1 cup (235 ml) milk, simply replace with 1 cup (235 ml) nondairy milk, such as soy or almond milk. Same rule applies for yogurt, sour cream, and butter. All you have to do is replace these items at a 1:1 ratio with their nondairy counterpart.

IF THE ORIGINAL RECIPE CALLS FOR...	REPLACE WITH...
1 cup (235 ml) buttermilk	• Combine 1 tablespoon (15 ml) fresh lemon juice or vinegar, such as apple cider or white balsamic, with 1 cup (235 ml) unsweetened soymilk
1 cup (235 ml) cow's or goat's milk	• 1 cup (235 ml) soymilk • 1 cup (235 ml) almond milk • 1 cup (235 ml) hemp milk • 1 cup (235 ml) rice milk • 1 cup (235 ml) coconut milk for drinking, such as So Delicious
1 cup (224 g) dairy butter	• 1 cup (224 g) nonhydrogenated, nondairy butter, such as Earth Balance • 1 cup (224 g) coconut oil • ¾ cup (168 g) vegetable shortening
1 cup (235 ml) heavy cream	• 1 cup (235 ml) soy creamer • 1 cup (235 ml) full-fat unsweetened coconut milk, or coconut cream
2 scoops ice cream	• 2 scoops soy ice cream • 2 scoops rice ice cream • 2 scoops coconut ice cream • 2 scoops Vanilla Latte Ice Cream (page 38) • 2 scoops Maple Orange Creamy Sorbet (page 38) • 2 scoops Pumpkin Ice Cream (page 40)
1 cup (240 g) sour cream	• 1 cup (240 g) nondairy sour cream, such as Tofutti or Follow Your Heart • 1 cup (240 g) Basic Tofu Sour Cream (page 21)
1 cup (240 g) yogurt	• 1 cup (240 g) coconut yogurt • 1 cup (240 g) rice yogurt • 1 cup (240 g) soy yogurt • 1 cup (240 g) Basic Homemade Sorta Yogurt (page 20)

VEGANIZED!: SAMPLE RECIPE

Have a look at the following traditional recipe for an example of how we would replace the dairy and other nonvegan ingredients:

BROWNIE PUDDING CAKE

This pudding-like chocolate cake has been adapted from the *Better Homes and Gardens New Cook Book*.

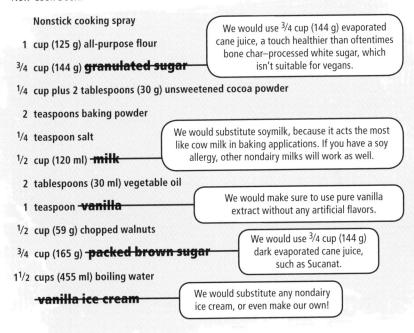

Nonstick cooking spray

1 cup (125 g) all-purpose flour

$^3/_4$ cup (144 g) ~~granulated sugar~~

> We would use $^3/_4$ cup (144 g) evaporated cane juice, a touch healthier than oftentimes bone char–processed white sugar, which isn't suitable for vegans.

$^1/_4$ cup plus 2 tablespoons (30 g) unsweetened cocoa powder

2 teaspoons baking powder

$^1/_4$ teaspoon salt

$^1/_2$ cup (120 ml) ~~milk~~

> We would substitute soymilk, because it acts the most like cow milk in baking applications. If you have a soy allergy, other nondairy milks will work as well.

2 tablespoons (30 ml) vegetable oil

1 teaspoon ~~vanilla~~

> We would make sure to use pure vanilla extract without any artificial flavors.

$^1/_2$ cup (59 g) chopped walnuts

$^3/_4$ cup (165 g) ~~packed brown sugar~~

> We would use $^3/_4$ cup (144 g) dark evaporated cane juice, such as Sucanat.

$1^1/_2$ cups (455 ml) boiling water

~~vanilla ice cream~~

> We would substitute any nondairy ice cream, or even make our own!

PREHEAT OVEN to 350°F (180°C, or gas mark 4). Lightly coat an 8-inch (20 cm) square baking dish with cooking spray.

In a medium bowl, stir together flour, evaporated cane juice, 2 tablespoons (10 g) of cocoa powder, baking powder, and salt.

Combine milk, oil, and vanilla in a small bowl. Fold wet ingredients into dry, being careful not to overmix. Fold in walnuts.

Pour batter evenly into prepared baking dish.

In a small bowl, stir together Sucanat and remaining $^1/_4$ cup (20 g) cocoa powder.

Combine with boiling water and slowly pour mixture over batter.

Bake for 40 minutes. Place on a wire rack for about 30 minutes.

Scoop out while still warm and top with vanilla nondairy ice cream.

YIELD: 6 to 8 servings

Finding Dairy Substitutes at the Store

Finding nondairy versions of old favorites has never been easier. With the ever-growing popularity of veganism, as well as the prevalence of dairy allergies, mainstream grocers have expanded their offerings of these products.

From milk to butter to sour cream to yogurt and even kefir, there is an amazing amount of nondairy substitutes available at your local market.

Right alongside the cow's milk, it is now common to find soymilk, almond milk, and rice milk. These can also be found in aseptic packages for a longer shelf life. Just remember to shake these milks as ingredients have a tendency to settle!

Next to the Ben and Jerry's, you can easily find delicious dairy-free ice creams made from soy, rice, and coconut milk. Sitting right alongside the butter, you will find dairy-free margarines both in tubs and in sticks.

If your local supermarket doesn't carry these items, our advice is to ask for them. Most grocers are more than willing to place special orders for customers, and if the items sell well, they will start ordering more!

Almost all store-bought nondairy milk will keep well for seven to ten days once opened, when stored in the fridge.

Almond milk: Tasty and creamy, rich in protein, this is a favorite in our kitchens and we use it a lot for both cooking and baking. We like it unsweetened, but it comes in a number of flavors, too.

Coconut milk: Both in its full-fat form and lighter versions, coconut milk adds creamy depth of flavor to many savory and sweet dishes. Recently made available in carton form specifically for drinking, this milk also tastes great over cereal or all by itself in a glass.

Hemp milk: A bit of an acquired taste; we prefer it in its chocolate version when we are enjoying it as a beverage. Its protein levels fall between those of soy and rice milks. It's good for some cooking uses and fabulous in lattes.

Nondairy butter or **margarine:** There are lots of brands on the market, and many of them still contain dairy and nasty hydrogenated fats. Be careful when shopping and keep an eye out for sneaky ingredients like whey and casein. We prefer Earth Balance both for its taste and performance.

Nondairy creamer: Thick and creamy, it makes a wonderful addition to your morning cuppa, as well as a nice enhancement for savory sauces, ice creams, and other desserts. We like to use the unsweetened kind when cooking and baking, as it is far more versatile than the flavored versions.

When it comes to heavier types of creams, we hear great things about MimicCreme, Soyatoo!, and Oatly. These brands can be found in your local health-food store or online.

Nondairy sour cream: Several brands are available that taste so rich they will have you doing double takes as to whether or not it's the real thing. Look for brands such as Tofutti or Follow Your Heart that do not contain hydrogenated fats.

Nondairy yogurts: Now available in many varieties made from soy, rice, and even coconut milk. Many flavors are available for snacking, baking, and cooking. Look for brands that contain live active cultures to get the full benefits of the creamy stuff.

Nut butters: Peanut, cashew, almond, hazelnut, pistachio—so many possibilities, so little time! Try to purchase organic as often as you can, especially with regards to peanut butter, as one of our favorite beans tends to contain a lot of pesticides.

Make your own butters using a high-power blender such as a Vita-Mix, or toast whatever nut you want to turn into butter, and use a coffee grinder. The results are quite amazing, even though cleaning up can be a bit of a pain.

Powdered milk: Soy, rice, and coconut milk powders are available if you only use milk occasionally. Be careful with the latter, however, as some brands use casein, for example, as an additive.

Rice milk: Thinner than most other nondairy milks, but free of most allergens, making it a perfect addition to households where allergies are an issue. It's rather low in protein and doesn't perform very well in most cooking applications as it is not as rich and creamy as nondairy milks such as coconut or soymilk. We feel it's best enjoyed as a beverage or with breakfast cereal.

Soymilk: The nondairy milk that is most commonly found in stores worldwide. Be sure to check for ingredients, as they are not all created equal. Look for those enriched in calcium to get the most out of your drinking experience, and make sure that the usually added vitamin D is D_2, not D_3, as the latter is a by-product from sheep.

It is possible to find a flavor to suit any palate or cooking and baking need: unsweetened, chocolate, vanilla, plain, and even crazier flavors such as strawberry-banana, soy nog, pumpkin spice, or peppermint chocolate.

Be sure to do a taste test to find the soymilk you like best.

Vegetable shortening: Good as a replacement for butter in baked goods. The general rule to replace 1 cup (224 g) butter is to use 3/4 cup (168 g) vegetable shortening. Look for a brand that uses nonhydrogenated oils and is of course free of animal products. Again, we favor Earth Balance both for its taste and performance.

Making Dairy Substitutes at Home

Purchasing nondairy milks can get expensive, so if you have the time, why not make your own? You'll know exactly where the ingredients come from and therefore won't have to worry about potential cross-contamination.

You will find that it is easy to get creative and sub your favorite nuts, seeds, grains, and flavorings for the ones we chose in the homemade milk recipes included here. The texture will depend on the amount of water you choose to use.

Costly equipment isn't even a requirement, as your good old blender or juicer will do a proper job of it. All you need is a fine-mesh strainer and cheesecloth, or one of the nut milk bags that are available for purchase online or at health-food stores. They're easy to clean and a bit less wasteful than cheesecloth.

If you have the opportunity, try to buy soybeans, nuts, seeds, rice, and other grains in bulk to lower your cost and packaging waste. Also, if you can afford organic, it is always best.

USE NONDAIRY BUTTERMILK

Combine 1 tablespoon (15 ml) fresh lemon juice or vinegar, such as apple cider or white balsamic, with 1 cup (235 ml) unsweetened nondairy milk (soy works best). Let stand a few minutes to let the milk curdle.

USE COCONUT MILK

Combine equal parts unsweetened dried coconut and water in a medium saucepan. Bring to a boil. Remove from heat, and let sit for one hour. Strain mixture through a fine-mesh sieve lined with cheesecloth, or use a blender on high speed and then strain.

USE NONDAIRY EVAPORATED MILK

Mixing together $2/3$ cup (122 g) dry soy or rice milk powder with $3/4$ cup (180 ml) water will yield 1 cup (235 ml) evaporated milk. Use at a 1:1 ratio for evaporated milk called for in a recipe.

USE NONDAIRY HALF-AND-HALF

Combine $1/2$ cup (120 ml) full-fat coconut milk with $1/2$ cup (120 ml) any unsweetened nondairy milk. Yields 1 cup (235 ml) nondairy half-and-half.

USE INSTANT NONDAIRY COFFEE CREAMER

Mix 1 cup (184 g) soy or rice milk powder with 1 tablespoon (6 g) vanilla powder and $1/4$ cup (30 g) powdered sugar for an instant creamy boost for your coffee.

USE SWEETENED CONDENSED SOYMILK

Bring 3 cups (705 ml) soymilk and 1 cup (192 g) evaporated cane juice to a simmer over low heat and simmer until liquid reduces to 1 cup (235 ml). Strain any lumps or solids before using.

The Recipes: Dairy Based Recipes ... without the Dairy!

First we'll take you back to basics with simple tried and true recipes for dairy substitutes that you can make at home. Then, we will showcase delicious dishes using homemade or store-bought dairy substitutes.

BASIC OLIVE OIL BUTTER

 Gluten Free   Quick and Easy   Wheat Free

This fast and easy spread makes an easy task of whipping up homemade "butter." Though it doesn't quite perform like the store-bought stuff when cooking and baking, it certainly tastes wonderful on toast or a bagel, or when used to make garlic bread or over hot corn bread or muffins.

1/2 cup (120 ml) olive oil
6 ounces (170 g) silken tofu
1 teaspoon maca powder
1/2 teaspoon sea salt
1/2 teaspoon xanthan gum

USING A BLENDER, purée all ingredients until silky and whipped.
    Store in an airtight container in the fridge for up to 2 weeks.

YIELD: 1 heaping cup (about 200 g)

Variations: Substitute canola (or any mild-flavored vegetable oil) for the olive oil and add 3 tablespoons (63 g) agave to make whipped honey butter. Tastes great with corn bread.
    Add 1 tablespoon (15 g) minced garlic, 1 teaspoon dried basil, and 1/4 teaspoon paprika to make a spread that's perfect for garlic bread.

BASIC HOMEMADE SORTA YOGURT

 Gluten Free Low Fat Quick and Easy Wheat Free

While this is not your typical yogurt packed with live cultures and other goodies, we find it is a handy recipe since it does exactly the same trick store-bought yogurt does in baked goods, while being cheaper and more readily available.

The amount of cornstarch used will depend on the thickness and quality of the milk: 1 tablespoon is enough for soymilk, 2 tablespoons are necessary for coconut milk.

For a no-cook version, replace the cornstarch with ½ teaspoon xanthan or guar gum and blend until thickened.

    1   cup (235 ml) full-fat soy or coconut milk
    1   to 2 tablespoons (15 to 30 ml) fresh lemon juice (1 is enough to curdle milk,
        2 brings a typical yogurt sourness)
    2   tablespoons sweetener of choice (24 g raw sugar, 42 g agave nectar . . .)
    1   to 2 tablespoons (8 to 16 g) cornstarch, depending on thickness of milk
        Pinch fine sea salt
    1   teaspoon pure vanilla extract, optional
        Lemon zest, optional
        Fresh fruit, optional

COMBINE MILK and lemon juice in a medium bowl, let curdle for 2 minutes.

Add sweetener, cornstarch, and salt.

Using an immersion blender, blend ingredients for one minute.

Heat in microwave for 1 minute for soymilk, 2 minutes for coconut milk. Throughout cooking, keep a close eye to make sure the mixture doesn't bubble up. Blend again.

Heat in microwave for 1 minute for soymilk, 2 minutes for coconut milk. Blend again or simply stir with a fork.

Heat in microwave for 1 more minute: The top of the mixture will look like a thin pudding. Stir in optional flavorings or fruit.

Store in fridge to thicken. Stir again before using.

Enjoy chilled within 4 days of preparation.

YIELD: 1½ cups (375 g)

BASIC TOFU SOUR CREAM

 Corn Free   Quick and Easy   Wheat Free

Even though you can buy premade nondairy sour cream, sometimes it's easier to make your own!

- 7 ounces (198 g) extra-firm tofu, pressed and drained
- 1/4 cup (28 g) raw cashews, ground into a fine powder
- 1 tablespoon (15 ml) white rice vinegar
- 1 tablespoon (15 ml) lemon or lime juice
- 1 tablespoon (18 g) white miso
- 1 tablespoon (15 ml) mild-flavored vegetable oil

PLACE ALL INGREDIENTS in a blender or food processor and blend until very, very smooth and creamy. Keep refrigerated in an airtight container until ready to use. Should last up to a week.

YIELD: 1¹/₂ cups (355 ml)

BASIC PEANUT MILK

 Corn Free   Gluten free   Soy Free   Wheat Free

This takes our love for peanuts to a whole new level by turning them into a milk perfect for baking and cooking and leaving behind a pulp that replaces tofu in baking quite beautifully. Adjust the amount of water and sweetener to obtain the perfect thickness, creaminess, and flavor you prefer. Check out our Pulp Not Fiction Muffins (page 75) that make good use of this milk and its leftover pulp.

- 2 cups (304 g) finely chopped peanuts
- 1/4 teaspoon fine sea salt
- 2 tablespoons (24 g) Sucanat
- 1/4 teaspoon ground cinnamon, optional
- 3 cups (705 ml) filtered water

PLACE PEANUTS, salt, and Sucanat in a large frying pan. Toast on medium heat, stirring constantly, until fragrant and lightly brown, about 2 minutes.

Add cinnamon, if using, and water.

Spoon mixture into blender and blend until smooth.

Strain through a fine-mesh sieve, lined with cheesecloth, or use nut milk bags. Do not discard the peanut pulp!

Store in fridge. Stir or shake before using.

YIELD: 2 cups (470 ml)

BASIC VANILLA CASHEW MILK

 Corn Free   Gluten Free   Soy Free   Wheat Free

Unlike a lot of other nut milks, this one needs no straining. In addition, you can make this without soaking the nuts if you are in a hurry (just make sure to blend it a little longer). Cashew milk is a little thicker than other nut milks, so it lends itself nicely to sweet applications and tastes amazing in coffee. Feel free to leave out the vanilla and agave if you want to use this in a savory dish (wonderful in curries!).

5 ounces (140 g, about 1 cup) raw cashews
2 tablespoons (42 g) agave nectar
1 tablespoon (15 ml) pure vanilla extract
2 cups (470 ml) filtered water

SOAK THE CASHEWS in water overnight in the fridge. Rinse in clean water and drain.
   Add soaked nuts, agave, and vanilla to a blender and add water slowly as you purée on high until milky, about 3 minutes, or until there is no more "grit."

YIELD: 2¹/2 cups (590 ml)

Note: For thinner milk, increase the amount of water to get your desired consistency. For a thicker cream, reduce the amount of water to reach your desired thickness.

WHITE OR MILK CHOCOLATE BAR

 Dairy Sub: rice or soymilk powder for non-fat dry milk powder

Use your homemade chocolate in cookies by breaking it into small chunks to replace chocolate chips, or create specialty chocolates of your own! You can add different flavored extracts, even loose tea leaves, food-grade lavender petals, and lemon peel to create one-of-a-kind confections for gift giving or personal enjoyment.

The optional soy lecithin granules, found in the supplement section of your local health-food store or online, are there to help promote bonding between the ingredients. Feel free to leave them out if you don't have easy access to them, but make sure to completely dissolve and mix all other ingredients.

FOR WHITE CHOCOLATE:

- 3 ounces (84 g) food-grade raw cacao butter (not for use as body cream)
- 1/2 cup (60 g) powdered sugar
- 1 tablespoon (12 g) rice milk or soymilk powder
- 1 teaspoon pure vanilla powder
- 1/4 teaspoon soy lecithin granules, optional

FOR MILK CHOCOLATE:

- 2 ounces (56 g) food-grade raw cacao butter
- 2 tablespoons (23 g) rice milk or soymilk powder
- 2 tablespoons (10 g) unsweetened cocoa powder
- 1/4 cup (30 g) powdered sugar
- 1 teaspoon pure vanilla powder
- 1 tablespoon (14 g) soy lecithin granules, optional

IN A SMALL SAUCEPAN, melt cacao butter on very low heat. Once liquefied, add in remaining ingredients and continue to stir until everything is completely dissolved.

Pour into a mold (a loaf pan works nicely) and refrigerate for about 1 hour to harden.

YIELD: One 4-ounce (112 g) bar

BERRY SUISSE BIRCHER MUESLI

 Dairy Sub: nondairy yogurt and nondairy milk

 Corn Free   **Wheat Free**

If you feel like starting your day on the right foot, try this version of the typical Swiss breakfast. Prepare it the night before and enjoy first thing in the morning.

- 2 **3.5-ounce (100 g) packages store-bought roasted chestnuts**
- 3/4 **cup (180 ml) apple cider**
- 1/2 **cup (120 g) nondairy yogurt**
- 1/2 **cup (160 g) all-fruit berry spread**
- 3/4 **cup (180 ml) nondairy milk, more if needed**
- 2 **tablespoons (42 g) agave nectar**
- 1 **teaspoon pure vanilla extract**
-    **Pinch ground cinnamon**
-    **Pinch sea salt**
- 1/3 **cup (40 g) dried berries**
- 1 **cup (80 g) old-fashioned oats**
- 3/4 **cup (113 g) fresh blackberries and raspberries**
-    **Handful sliced almonds**

PLACE CHESTNUTS and 1/2 cup (120 ml) of cider in a saucepan. Bring to a boil, lower heat, and simmer for 6 minutes, or until the chestnuts are soft enough to mash coarsely and most of the liquid is gone. Set aside to cool.

Using a blender, blend yogurt, berry spread, 1/2 cup (120 ml) of milk, remaining 1/4 cup (60 ml) cider, agave, vanilla, cinnamon, and salt.

Stir in cooled chestnuts, dried berries, and oats. The mixture will look thinner than it should, but the oats will use the extra moisture to thicken up. If you find that the preparation is actually not thin enough, add more cider or milk.

Cover and let sit overnight in the fridge to let the oats "cook."

Add remaining 1/4 cup (60 ml) milk if the mixture is too thick. Adjust sweetness to taste. Serve with fresh berries. Decorate each serving with a sprinkle of almonds. Keeps for up to 4 days in the fridge.

YIELD: 4 cups (996 g)

BASIC MAPLE CREAM

 Corn Free   Gluten Free   Wheat Free

Need something lusciously sweet and creamy to accompany your waffles and other sweet treats? This cream tastes amazing served with Carrot Cake Waffles (page 80)!

1 cup (117 g) walnut halves or (99 g) pecan halves
3/4 cup (180 ml) pure maple syrup
  Pinch fine sea salt
2 tablespoons (28 g) nondairy butter
2 teaspoons maple extract
1/2 cup (96 g) Sucanat
12 ounces (340 g) extra-firm silken tofu
2 teaspoons pure vanilla extract
1 teaspoon ground cinnamon, optional

Note: Got leftover Maple Cream? Proceed to the recipe below!

COMBINE NUTS, 1/4 cup (60 ml) syrup, salt, and butter in small saucepan. Cook over medium heat for a couple of minutes, until the nuts become fragrant and the mixture thickens up a little.

Using an immersion blender or a food processor, blend all ingredients until perfectly smooth.

Chill for at least 4 hours to thicken. Use in waffles or as a topping for cakes. Keep refrigerated for up to a week.

YIELD: 2 1/2 cups (590 ml)

BASIC CHOCOLATE MAPLE DESSERT

 Dairy Sub: just naturally dairy free!

1 cup (235 ml) Basic Maple Cream (above)
9 ounces (255 g) semisweet chocolate chips
12 ounces (340 g) silken soft tofu
2 tablespoons (30 ml) pure maple syrup

COMBINE MAPLE CREAM and chocolate chips in a medium bowl. Heat in the microwave for about 1 minute, or until the chips melt when stirring them with the maple cream.

Combine with tofu and maple syrup. Using a blender, purée until perfectly smooth.

Serve chilled or frozen in small dishes. Thaw for a few minutes before enjoying.

YIELD: 6 to 8 servings

SAVORY CHUTNEY MUFFINS

 Dairy Sub: nondairy milk (almond)

 Soy Free

Tahini and sesame oil pair up beautifully in these tasty muffins to create a rich and palatable muffin that will be perfect to buddy up with your Indian-style dishes.

Because the texture of both tahini and chutneys can vary, you might have to add more milk or water if the batter is too thick.

Nonstick cooking spray

$1/2$ cup (144 g) spicy mango chutney

$1/4$ cup (64 g) tahini

1 tablespoon (15 ml) sesame oil (not toasted)

2 teaspoons ground coriander

$1/2$ teaspoon red pepper flakes, plus more to sprinkle on top

1 cup (235 ml) unsweetened almond milk

$1^3/4$ cups (210 g) whole wheat pastry flour

2 teaspoons baking powder

1 teaspoon fine sea salt

1 teaspoon cumin seeds

PREHEAT OVEN to 350°F (180°C, or gas mark 4). Lightly grease a jumbo muffin tin with cooking spray.

In a large bowl, combine chutney, tahini, and sesame oil until emulsified. Add coriander, red pepper flakes, and milk. Sift flour, baking powder, and salt on top.

Fold dry ingredients into wet, being careful not to overmix. Add extra milk if needed. Divide batter equally into muffin cups. Sprinkle cumin seeds and pepper flakes on top of each muffin.

Bake for 40 minutes, or until a toothpick inserted into center of muffin comes out clean. Remove from the pan and let cool on a wire rack.

YIELD: 6 jumbo muffins

Note: If you don't have a jumbo muffin tin, bake batter in a regular muffin pan for 20 minutes. You'll get about 12 muffins.

BUTTERY SAVORY QUICK BREAD

 Dairy Sub: nondairy butter and nondairy milk

This is the type of rich, buttery, dense quick bread that just belongs with soups and stews. It contains quite a large amount of nondairy butter, which makes things even easier by cutting one step, since you won't have to spread anything on it before enjoying every bite.

Nonstick cooking spray

4 ounces (112 g) nondairy butter

4 ounces (112 g) unsweetened applesauce

1 cup (235 ml) vegan beer or nondairy milk

1 cup (80 g) old-fashioned or quick oats, finely ground

1 cup (120 g) white whole wheat flour

1 cup (120 g) bread flour

1 tablespoon (12 g) baking powder

1 teaspoon fine sea salt

Handful sunflower seeds or sliced almonds to sprinkle on top

PREHEAT OVEN to 375°F (190°C, or gas mark 5). Lightly coat an 8 x 4-inch (20 x 10 cm) loaf pan with cooking spray.

Combine butter, applesauce, and beer or milk in a microwave-safe bowl, and heat for 40 seconds. Stir to melt the butter. Set aside to cool.

In a large bowl, whisk together ground oats, flours, baking powder, and salt.

Fold wet ingredients into dry, being careful not to overmix.

Place batter into prepared pan. Sprinkle a generous handful of seeds or almonds on top, pressing down gently to make sure they adhere to the batter.

Bake for 55 minutes, or until a toothpick inserted in the center comes out clean. Loosely cover with a piece of foil if bread starts to brown too quickly.

Remove from oven and cool on a wire rack.

YIELD: One 8-inch (20 cm) loaf

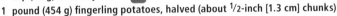

PAPET VAUDOIS

 Dairy Sub: nondairy butter and nondairy milk or creamer

 Nut Free  **Wheat Free**

Papet vaudois is a rich, traditional dish of leeks with sausage from the French-speaking region of Switzerland. It is usually made with bacon, sausage, cream, and dairy butter. What a treat it is to find once again that the animal-friendly version is just as good (perhaps even better) than the so-called real thing!

Serve with your favorite sauerkraut-, onion-, or beer-based vegan bratwursts.

¹/₄ cup (56 g) nondairy butter
1 pound (454 g) fingerling potatoes, halved (about ¹/₂-inch [1.3 cm] chunks)
3 cups (267 g) leeks, white and light green parts only, thoroughly cleaned, cut in 1-inch (2.5 cm) chunks
1 large onion, cut in 1-inch (2.5 cm) chunks
¹/₂ teaspoon fine sea salt
¹/₄ teaspoon ground nutmeg
   White pepper to taste
¹/₃ cup (80 ml) vegetable broth
¹/₃ cup (80 ml) dry white wine
1 teaspoon liquid smoke
¹/₃ cup (80 ml) unsweetened nondairy milk or creamer

MELT 3 TABLESPOONS (42 g) of butter over medium-high heat in a large saucepan or skillet. Add potatoes, leeks, onion, and salt. Sauté for 10 minutes, until the potatoes just start to brown.

Add nutmeg, pepper, broth, wine, and liquid smoke. Bring to a boil.

Cover and cook over medium heat for 20 minutes, or until the potatoes are tender. Remove lid, add remaining 1 tablespoon (14 g) butter along with milk or creamer. Cook until very little liquid is left.

YIELD: 4 servings

GRANOLA BISCUITS

 Dairy Sub: nondairy butter and nondairy milk

Flaky, buttery, and delicate—who needs real butter when you can have these for breakfast instead of the heavily processed biscuits popped from a can?

1¹/₂ cups (110 g) granola

1¹/₂ cups (188 g) all-purpose or whole wheat pastry flour

2 teaspoons baking powder

¹/₂ teaspoon fine sea salt

¹/₂ teaspoon ground cinnamon

¹/₄ cup (56 g) cold nondairy butter, cut into small pieces

2 tablespoons (30 ml) pure maple syrup

¹/₄ cup to ¹/₄ cup plus 2 tablespoons (60 to 90 ml) cold nondairy milk

PREHEAT OVEN to 375°F (190°C, or gas mark 5). Line a baking sheet with parchment paper on a silicone baking mat.

Combine granola, flour, baking powder, salt, cinnamon, butter, and syrup in food processor. Pulse a few times until combined.

Add milk until moist dough forms.

Divide dough into 6 equal portions of one heaping ¹/₄ cup (77 g) per biscuit.

Pat dough to shape into round, 1-inch (2.5 cm)-thick biscuits.

Bake for 14 to 16 minutes, until biscuits are light golden on top and at the bottom.

Let cool on a wire rack.

YIELD: 6 biscuits

PUMPKIN-DRESSED SLAW WITH CHIPOTLE WALNUTS

 Dairy Sub: vegan mayonnaise (for creaminess)

 Corn Free   **Wheat Free**

Use this creamy and comforting slaw as a sandwich filling or enjoy it on its own.
No matter what, be sure to combine with the walnuts: The two just belong together!

FOR CHIPOTLE WALNUTS:
- 1 cup (117 g) walnut halves
- 1 tablespoon (15 g) adobo sauce from a can,
  including 1/2 teaspoon finely chopped chipotle pepper
- 1/2 teaspoon ground cumin
- 1/2 teaspoon unsweetened cocoa powder
- 1/4 teaspoon fine sea salt
- 1 teaspoon agave nectar

FOR PUMPKIN-DRESSED SLAW:
- 5 tablespoons (70 g) vegan mayonnaise
- 1/2 cup (122 g) pumpkin purée
- 1/2 teaspoon fine sea salt
- 1 teaspoon onion powder
- 2 tablespoons (30 ml) apple cider vinegar
- Pepper to taste
- 1 teaspoon Dijon mustard
- 10 ounces (283 g) shredded green cabbage

TO MAKE THE CHIPOTLE WALNUTS: Preheat oven to 325°F (170°C, or gas mark 3).
Combine all ingredients and spread onto a baking sheet; toast for 10 minutes, stirring
once halfway through.
Let cool on sheet.

TO MAKE THE SLAW: Combine all ingredients except cabbage, until emulsified.
Add cabbage and stir well. Chill for 1 hour to let flavors meld.
Serve with walnuts.

YIELD: 4 servings

PEAR AND CAULIFLOWER CURRY

 Dairy Sub: coconut oil and coconut milk

 Corn Free   Gluten Free   Soy Free   Wheat Free

This easy and spicy recipe is so low-key you won't know what to do with the free time it'll give you. We love to pair it with Carrot Ginger Burgers (page 114).

3 tablespoons (42 ml) melted coconut oil
1 tablespoon (6 g) garam masala
1 teaspoon ground cumin
1/2 teaspoon ground ginger
1/2 teaspoon ground coriander
1/2 teaspoon brown mustard seeds
1/2 teaspoon red chili flakes
1 teaspoon turmeric
1 teaspoon coarse sea salt
1 small head cauliflower, broken into small florets
2 firm pears, quartered and cored
2 cloves garlic, peeled and minced
1/3 cup (53 g) chopped yellow onion
1/4 cup (60 ml) coconut milk
1/3 cup (45 g) dry-roasted salted cashews
1/4 cup (15 g) chopped fresh cilantro or parsley

PREHEAT OVEN to 400°F (200°C, or gas mark 6). Combine all ingredients in a 9 x13-inch (23 x 32 cm) baking dish.

Roast for 30 minutes, stirring once halfway through. Turn oven off and leave dish in the oven for another 15 minutes, or until the cauliflower is fork tender without being mushy.

Sprinkle with cashews and cilantro or parsley before serving.

YIELD: 4 servings

Note: Try serving this curry atop quinoa or rice, or alongside some meatless burgers.

CREAMY BUTTERNUT PASTA BAKE

 Dairy Sub: almonds, nondairy milk, and vegan mayonnaise (for creaminess)

Pasta is a surefire way to get your energy roaring, and when you dress it up with flavor-rich and nutritious ingredients that more than match the traditional cheesiness you're accustomed to? There truly is nothing better than that. Serve with Curly Mustard Greens (page 262).

- 5 cups (1 kg) cubed butternut squash
- 2 tablespoons (30 ml) extra-virgin olive oil
- 1 pound (454 g) rotini pasta
- 1 cup (160 g) dry-roasted almonds 🔲
- 2 tablespoons (30 ml) adobo sauce, from a chipotle can
- 1 tablespoon (15 g) chopped canned chipotle
- ¹/₂ teaspoon fine sea salt
- 2 teaspoons ground coriander
- 1 teaspoon ground cumin
- 3 cloves garlic, peeled and minced
- ¹/₄ cup (40 g) chopped onion
- 2 tablespoons (30 ml) triple sec or orange juice
- 1¹/₄ cups (295 ml) unsweetened nondairy milk 🔲
- ¹/₄ cup (56 g) vegan mayonnaise or sour cream 🔲
- 2 tablespoons (30 ml) fresh lemon juice

TO ROAST SQUASH, preheat oven to 400°F (200°C, or gas mark 6). Toss cubes with olive oil and spread on baking sheet. Cook for 45 minutes, or until tender. Set aside.

Cook pasta according to package directions. Drain and set aside.

In a food processor, blend almonds into a fine powder.

In a large saucepan, combine squash, adobo sauce, chipotle, salt, coriander, cumin, garlic, and onion. Cook over medium heat for 6 minutes, until the onions are translucent.

Add triple sec or juice; simmer for 2 minutes.

Remove from heat. Transfer mixture to food processor with almonds, milk, mayonnaise or sour cream, and lemon juice. Blend until smooth.

Preheat oven to 375°F (190°C, or gas mark 5). Combine cooked pasta with sauce in a 9 x 13-inch (23 x 33 cm) baking dish. Press down evenly.

Bake for 20 minutes, or until golden brown. Let stand for 15 minutes before serving. Tastes even better the next day. Freezes well.

YIELD: 6 servings

PUMPKIN BISQUE

 Dairy Sub: nondairy milk

 Quick and Easy

This flavorful bisque is ready in mere minutes. Great as a snack or as a more substantial meal served alongside crusty bread.

1 can (15 ounces, or 425 g) pumpkin purée

1¹/₂ tablespoons (39 g) yeast extract spread, such as Marmite

3 tablespoons (48 g) natural peanut butter

2 cloves garlic, peeled and chopped

3 tablespoons (30 g) finely chopped onion

2 teaspoons garam masala
Pepper

¹/₂ teaspoon cayenne pepper

1 teaspoon vegan Worcestershire sauce

2 cups (470 ml) unsweetened nondairy milk

2 cups (470 ml) water or vegetable broth

2 teaspoons pure maple syrup

COMBINE ALL INGREDIENTS, except maple syrup, in a soup pot.

Bring to a low boil, lower heat, and simmer for 10 minutes, whisking occasionally. Add maple syrup.

Using a blender, purée to desired texture, and serve.

YIELD: 4 servings

TRIPLE OLIVE SPREAD

 Dairy Sub: silken tofu (for creaminess)

 Corn Free

 Quick and Easy

 Wheat Free

Delicious as a sandwich spread but also great alternative to pesto on pasta or baked potatoes.

12 ounces (340 g) soft silken tofu, drained
1/2 cup (67 g) pitted black olives
8 pitted green olives
1 teaspoon capers, without brine
2 tablespoons (30 ml) white balsamic vinegar
1/2 cup (20 g) packed fresh basil leaves
3 tablespoons (45 ml) extra-virgin olive oil
Fine sea salt to taste
Pepper to taste
Pinch red pepper flakes to taste
Zest of 1/2 large lemon
1 large clove garlic, peeled and chopped
1/4 cup (29 g) toasted walnuts

USING A BLENDER, purée all ingredients until smooth. Store covered in the fridge. Keeps well for up to 4 days.

YIELD: 2 cups (555 g)

ROASTED TOMATO AIOLI

 Dairy Sub: vegan mayonnaise (for creaminess)

 Corn Free   **Wheat Free**

Serve this creamy concoction as a hot or cold dressing for 1 pound (454 g) cooked pasta, with baked potatoes, or as a sandwich spread.

FOR ROASTED TOMATOES:
1 pound (454 g) grape tomatoes
1 tablespoon (15 ml) extra-virgin olive oil
1/2 teaspoon fine salt
1/4 cup (40 g) chopped onion
3 cloves garlic, peeled and halved

FOR AIOLI:
1/2 cup (112 g) vegan mayonnaise
1 tablespoon (15 g) blended chopped canned chipotle pepper and adobo sauce
1 teaspoon paprika
1/2 teaspoon fine sea salt
2 teaspoons agave nectar
1/2 teaspoon ground cumin
2 cloves garlic, peeled and chopped
Pepper to taste

TO MAKE THE ROASTED TOMATOES:
Preheat oven to 400°F (200°C, or gas mark 6).
    Combine all ingredients in an 8-inch (20 cm) square baking pan.
    Roast for 20 minutes, stirring once halfway through, until the tomatoes look like deflated tires. Remove from oven and set aside.

TO MAKE THE AIOLI: Combine all ingredients and roasted tomatoes in a food processor or blender. Blend until smooth.
    Serve as is with pasta or baked potatoes, but chill to thicken before using as a sandwich spread.

YIELD: 2 3/4 cups (605 g)

CHOCOLATE PUDDING

 Dairy Sub: nondairy milk
(Chocolate soy)

 Gluten Free   **Nut Free**

 Wheat Free

Feel free to add various extracts to this delicious dessert, such as peppermint, almond, orange, or rum, to take your pudding from mighty tasty to simply irresistible. This can take the place of nondairy yogurts in recipes like the Walnut Chocolate Brownies (page 88).

2²/3 cups (630 ml) chocolate soymilk
 ¹/4 cup plus 2 tablespoons (60 g)
    Sucanat
    Pinch fine sea salt
 ¹/4 cup (20 g) unsweetened cocoa
    powder
 ¹/4 cup (32 g) cornstarch
  2 teaspoons pure vanilla extract

USING A BLENDER, blend all ingredients until smooth.
    Transfer into a large saucepan and cook over medium-high heat until thickened, about 6 minutes.
    Transfer to small dessert dishes, cover with plastic wrap. Chill overnight before enjoying.

YIELD: 2³/4 cups (700 g) or 6 to 8 servings

BERRY FRO YO

 Dairy Sub: nondairy yogurt

 Corn Free   **Gluten Free**

 Low Fat   **Wheat Free**

Nondairy yogurts are just as good as their dairy counterparts. Want proof? Try this frozen dessert!

3¹/4 cups (491 g) fresh or frozen berries,
    partially thawed
3   six-ounce (170 g) containers
    nondairy yogurt
    Juice of ½ a large lemon, about
    2 tablespoons (30 ml)
 ¹/4 cup (84 g) agave nectar
 ¹/2 teaspoon pure vanilla extract
 ¹/4 teaspoon pure hazelnut extract,
    optional
 ¹/8 teaspoon pure almond extract,
    optional

USING A BLENDER, blend all ingredients until smooth.
    Strain using a fine-mesh sieve, to remove pesky seeds.
    Place in ice-cream maker and follow manufacturer's instructions.
    Place in the freezer for an hour before enjoying for firmer frozen yogurt.

YIELD: 8 large scoops

VANILLA LATTE ICE CREAM

 Dairy Sub: vanilla cashew milk for cow's milk

 Gluten Free   **Soy Free**

This tastes amazing topped with caramel sauce, and makes good use of the Basic Vanilla Cashew Milk.

 **One recipe Basic Vanilla Cashew Milk (page 22) or 2^1/$_2$ cups (590 ml) nondairy milk**

- 2 **cups (470 ml) brewed coffee**
- 1 **cup (192 g) evaporated cane juice**
- ^{1}/$_4$ **cup (32 g) cornstarch combined with ^{3}/$_4$ cup (180 ml) water to make a slurry**
- 2 **tablespoons (30 ml) pure vanilla extract**

COMBINE MILK, coffee, and evaporated cane juice in a medium pot. Bring to a boil and immediately remove from heat.

Stir in cornstarch slurry to thicken. Stir in vanilla.

Place in the fridge to cool for a few hours before making it into ice cream. Follow the manufacturer's instructions on your ice cream maker.

YIELD: 7^1/$_4$ cups (36 ounces, or 1 kg)

MAPLE ORANGE CREAMY SORBET

 Dairy Sub: silken tofu (for creaminess)

This is one tasty and refreshing dessert wherein cookie pieces taste like little bits of caramel heaven.

- 1 **cup (235 ml) fresh orange juice**
- ^{3}/$_4$ **cup (144 g) raw sugar**
- 1 **cup (117 g) walnut halves**
- 12 **ounces (340 g) silken soft tofu**
- 1 **teaspoon pure vanilla extract**
- 1 **teaspoon maple extract**
  Pinch fine sea salt
- 10 **store-bought Speculoos cookies or 6 homemade Speculoos-Spiced Cookies (page 97), broken in small pieces**

COMBINE ALL INGREDIENTS (except cookies) in blender, and process until smooth. Chill for a few hours.

Place in ice-cream maker and follow manufacturer's instructions. Add cookie pieces about 5 minutes before the ice cream is done churning.

Place back into the freezer for a firmer consistency, or enjoy straightaway.

YIELD: 8 large scoops

FROZEN POTS DE CRÈME

 Dairy Sub: extra-firm or firm tofu (for creaminess and texture)

 Corn Free **Gluten Free** **Wheat Free**

Rich and bittersweet dessert, ahoy! The tofu adds a wonderful creaminess and texture to this treat, while the syrup, nectar, coffee, and chocolate give it maximal flavor.

1 cup (235 ml) Choco PB Syrup (page 232)

$^{1}/_{2}$ cup (168 g) agave nectar
Pinch fine sea salt

$^{3}/_{4}$ cup (180 ml) hot coffee

8 ounces (227 g) unsweetened baking chocolate, chopped

1 tablespoon (15 ml) pure vanilla extract

14 ounces (397 g) extra-firm or firm tofu, drained, squeezed and crumbled

COMBINE SYRUP, nectar, salt, and coffee in a microwave-safe dish, or in a saucepan on the stovetop. Heat for 1 minute, or until warmed through.

Add chocolate and stir into warm liquid until completely melted.

Using a blender or food processor, combine with vanilla and tofu, and blend until perfectly smooth, scraping sides occasionally. Transfer into 8 to 10 microwave-safe dessert dishes. Place into freezer.

Thaw for 30 minutes or microwave for 15 seconds before enjoying.

Alternatively, use an ice-cream maker, following manufacturer's instructions.

YIELD: 8 to 10 servings

PUMPKIN ICE CREAM

🐾 **Dairy Sub:** vegan cream cheese or vegan creamer (for creaminess and texture)

Move along, dairy ice creams: This one is rich, thick, and creamy, and it's not ashamed. If you have a hard time finding Speculoos cookies or don't have the time to make your own, use graham crackers instead; just be sure to check for nonvegan ingredients. Add a few toasted walnuts, if you want an even fancier treat.

12 (about 94 g) store-bought Speculoos cookies, plus 4 more as add-in

3/4 cup (144 g) Sucanat

1 1/4 cups (305 g) pumpkin purée

8 ounces (227 g) vegan cream cheese or thick vegan creamer 🐾

1/4 teaspoon lemon zest

2 teaspoons pure vanilla extract

2 teaspoon ground cinnamon

1/4 cup (44 g) semisweet chocolate chips

BLEND 12 COOKIES in food processor, until finely ground. Add remaining ingredients except the extra cookies and process until smooth. Chill for a few hours.

Place mixture in ice-cream maker and follow manufacturer's instructions. Stir in 4 extra cookies, broken into chunks, as add-ins during last 5 minutes of churning. Place back into the freezer for a firmer consistency, or enjoy right away.

YIELD: 8 large scoops

Variation: If you'd rather use toasted walnuts as an add-in, make 'em maple!

1 cup (117 g) walnut halves

1/4 cup (60 ml) pure maple syrup

Pinch fine sea salt

Preheat oven to 325°F (170°C, or gas mark 3). Combine all ingredients on a small baking sheet, and toast for 10 minutes, stirring once halfway through.

Remove from oven and let cool on sheet.

LEMON TART IN TWISTED GRAHAM CRACKER CRUST

 Dairy Sub: nondairy milk (soy) and nondairy butter

 Corn Free Nut Free

Pucker up, Buttercup! So you thought your days of enjoying tart and deliciously creamy lemon pies were over? We're happy to tell you, nothing could be further from the truth.

FOR FILLING:

2 cups (470 ml) plain or vanilla soymilk

2 teaspoons lemon zest

1/3 cup (80 ml) fresh lemon juice

1/2 cup (96 g) raw sugar

1 teaspoon pure vanilla extract

1/4 cup (30 g) all-purpose or light spelt flour

Pinch fine sea salt

FOR CRUST:

Nonstick cooking spray

1/2 cup (112 g) nondairy butter

1/4 cup (48 g) Sucanat

1 cup (140 g) whole spelt flour

1/4 cup (40 g) brown rice flour

1/2 teaspoon fine sea salt

1 teaspoon ground cinnamon

TO MAKE THE FILLING: Using an immersion blender, combine milk, zest, juice, sugar, vanilla, flour, and salt in a microwave-safe container.

Heat for 2 minutes, blend again. Heat another minute to thicken, set aside to cool.

TO MAKE THE CRUST: Preheat oven to 350°F (180°C, or gas mark 4). Lightly grease a 9-inch (23 cm) round baking pan with cooking spray.

Using an electric mixer, cream together butter and Sucanat.

Add flours, salt, and cinnamon. Mix until combined.

Crumble dough into prepared pan; press down with the palm of your hand or a rubber spatula, until the sides and bottom are covered with the crust. Lightly prick with a fork to avoid air bubbles.

Prebake crust for 15 minutes, turning once halfway through and making sure it doesn't burn. Remove from oven, place on a wire rack for 10 minutes.

Pour filling into crust. Bake for 30 minutes. Let cool completely before placing in the fridge, and let set overnight.

Enjoy chilled, within a day or two of preparation.

YIELD: 8 slices

Chapter 2
From Mouthwatering Mac to Brilliant Brie:
HOW TO SUBSTITUTE FOR CHEESE

CHEESE: THE LAST VESTIGE OF ANIMAL BY-PRODUCTS to keep vegetarians from making the final switch to veganism.

What is it about cheese that people love so much? It tastes good, it melts and it's gooey, it adds pizzazz to pizza, texture to salads, and what would a grilled cheese sandwich be without, well, cheese?

Be it sharp cheddar or a soft and creamy brie, we totally understand people who love the stuff. In this chapter we will introduce you to several alternatives that will help wean you off the casein-filled, artery-clogging, so-called real thing and replace that craving with nut, grain, bean, soy, and yeast-based cheese substitutes.

Consider the Facts

We've already told you about the addictive properties of dairy in chapter 1 (page 13). Cheese is made mostly from dairy, but in addition, many cheeses contain rennet, which is traditionally made using sliced up pieces of the dried-out stomachs of calves. Considering the fact that cheese is also loaded with saturated fats and sodium, these should be reasons enough to make anyone give up cheese.

Guidelines for Substituting for Cheese

Depending on the availability of ready-made nondairy cheeses in your area, replacing the cheese in traditional recipes can be a simple task, or one that requires a little bit of effort.

If you have access to good-tasting cheese substitutes, then you can just replace the cheese in the original recipe at a 1:1 ratio with a nondairy version.

If you are not fond of store-bought cheese alternatives or don't have easy access to them, you may have to be a little bit more creative. It can still be done! With ingredients such as cashews, walnuts, nutritional yeast, tofu, maca powder, agar, and miso, your handcrafted cheeses will not only taste delicious and offer irresistible texture, but they will also give you a powerful nutritious boost. These ingredients may seem a bit foreign at first, but you'll find that in no time, they will become indispensable to both your kitchen and taste buds, just like your favorite Parmesan and cheddar used to be.

IF THE ORIGINAL RECIPE CALLS FOR…	REPLACE WITH…
1 cup (112 g) shredded cheese	• 1 cup (112 g) nondairy shredded cheese, such as Daiya
1 slice cheese (for sandwiches)	• 1 slice Galaxy or Tofutti Cheese Slices (available in several flavors) • 1 slice Nutty Pepperjack (page 48)
1 cup (240 g) cream cheese	• 1 cup (240 g) nondairy cream cheese, such as Tofutti or Follow Your Heart
1 cup (235 ml) nacho cheese sauce	• 1 cup (235 ml) Nacho Queso (page 52)
1 cup (150 g) feta crumbles	• 1 cup (150 g) Tofu Feta (page 46)
1 cup (245 g) ricotta cheese	• 1 cup (245 g) Tofu Ricotta (page 47)

VEGANIZED!: SAMPLE RECIPE

Let's have a look at the following traditional recipe for an example in how we would replace the cheese and other nonvegan ingredients:

BAKED CHEESE GRITS

This hearty breakfast treat was adapted from the *Better Homes and Gardens New Cook Book.*

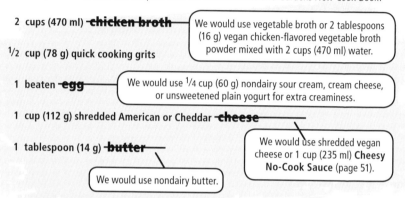

2 cups (470 ml) ~~chicken broth~~

We would use vegetable broth or 2 tablespoons (16 g) vegan chicken-flavored vegetable broth powder mixed with 2 cups (470 ml) water.

¹⁄₂ cup (78 g) quick cooking grits

1 beaten ~~egg~~

We would use ¹⁄₄ cup (60 g) nondairy sour cream, cream cheese, or unsweetened plain yogurt for extra creaminess.

1 cup (112 g) shredded American or Cheddar ~~cheese~~

We would use shredded vegan cheese or 1 cup (235 ml) **Cheesy No-Cook Sauce** (page 51).

1 tablespoon (14 g) ~~butter~~

We would use nondairy butter.

PREHEAT OVEN to 325°F (170°C, or gas mark 3). Have ready a 1-quart (0.95 L) casserole.
   In a saucepan, bring broth to a boil. Slowly add grits in a fine stream, whisking constantly.
   Remove from heat; stir in sour cream, cream cheese, or yogurt, along with shredded cheese or Cheese Sauce and nondairy butter until melted.
   Transfer mixture into casserole and bake for 25 to 30 minutes, or until a knife inserted into the center comes out clean.
   Let stand 5 minutes before serving.

YIELD: 4 servings

Finding Cheese Substitutes at the Store

Until recently, it was all but impossible to find store-bought cheese substitutes that not only tasted good but actually functioned like cheese in recipes. Several new brands have leaped into the marketplace with impressive results. Tofutti, Follow Your Heart, Galaxy, Teese, Sheese, and Dr-Cow have all made giant strides in offering good tasting cheese alternatives.

Our current favorite is Daiya, which is not only dairy free but also happens to be soy and gluten free. It has sprung up in specialty markets and pizza joints across the United States. It comes in white and yellow shreds as well as in solid block form. The stuff not only melts but stretches, too! The taste is mild and reminiscent of American processed cheeses. It works best when melted, tastes fantastic on pizza, in quesadillas, and grilled cheese sandwiches, and, of course, in everyone's favorite, macaroni and cheese.

Besides the cheese subs mentioned above, there are also a couple of really good cream cheese substitutes on the market. Both Tofutti and Follow Your Heart make excellent versions, which are pretty spot-on for the real thing and can be used interchangeably for the dairy version in recipes calling for cream cheese.

Making Cheese Substitutes at Home

The mere thought of going through the lengthy process of making dairy-based cheese at home isn't something most home cooks ponder, considering that cheese is so readily available at the store. Once you discover how easy it is, making vegan cheese at home is something you will not only come to consider—you'll actually look forward to it.

Like all true artisan cheeses, a homemade wheel of Baked Macadamia Nut Brie (page 48) or a pot of hot Fondue (page 53) will have you chomping at the bit to invite all your friends and family over for a vegan wine and cheese party!

The Recipes: Cheese and Cheese-Based Recipes . . . without the Cheese!

If the sometimes hard to find, and often costly store-bought vegan cheeses aren't really your thing, see how easy it can be to make your own incredible cheeses that can be devoured on their own or incorporated into other recipes.

Once again, we start with recipes for made-from-scratch cheeses and then offer up delectable dishes that will more than satisfy your cheese cravings—they may even make you ask why you ever liked the dairy stuff in the first place!

TOFU FETA

 Cheese sub: tofu for feta

 Corn Free   **Gluten Free**   **Nut Free**   **Wheat Free**

Make this a day ahead, as it is really important to let the flavors develop overnight. Use as you would any feta.

- 14 ounces (397 g) extra-firm tofu, drained and pressed
- 3 tablespoons (45 ml) extra-virgin olive oil
- 2 tablespoons (30 ml) lemon juice
- 1 tablespoon (2 g) dried basil, or 3 tablespoons (9 g) finely chopped fresh basil

    Salt and pepper to taste

CRUMBLE TOFU into a bowl until it resembles feta. Add in other ingredients and use hands to mix together.

Store in an airtight container in the fridge. Keeps well for at least a week.

Note: Add 2 tablespoons (17 g) capers for an extra-briny feta.

YIELD: About 2½ cups (590 ml)

TOFU RICOTTA

 Cheese sub: tofu for milk, cashews for creaminess, and nutritional yeast for cheesy flavor

 Corn Free Gluten Free

 Quick and Easy

 Wheat Free

Use as you would any ricotta, in lasagna, on pizza, in stuffed manicotti, etc.

- 14 ounces (397 g) extra-firm tofu, drained and pressed
- 1/4 cup (35 g) raw cashews, finely ground
- 1/4 cup (30 g) nutritional yeast
- 3 tablespoons (45 ml) extra-virgin olive oil
- 2 tablespoons (6 g) finely chopped fresh basil, or 1 tablespoon (2 g) dried basil
- Salt and pepper to taste

CRUMBLE TOFU into a bowl until it resembles ricotta.

Add in other ingredients and use hands to mix together.

Store in an airtight container in the fridge. Keeps well for at least a week.

YIELD: About 2¹/₂ cups (590 ml)

MARINATED BUFFALO-STYLE TOFU MOZZARELLA

 Cheese sub: silken tofu for fresh buffalo mozzarella

 Corn Free

 Quick and Easy

 Wheat Free

This simple preparation tastes great on toasted bread, crackers, as a pizza topping, in salads, sandwiches, and more.

- ¹/₂ cup (120 ml) extra-virgin olive oil
- ¹/₄ cup (60 g) julienne-cut sun-dried tomatoes, packed in oil
- 2 tablespoons (30 ml) balsamic vinegar
- 1 tablespoon (2 g) dried basil
- 1 tablespoon (15 g) minced garlic
- 1 teaspoon fennel seed
- ¹/₂ teaspoon sea salt
- ¹/₄ teaspoon fresh cracked black pepper
- 14 ounces (397 g) silken tofu, drained but not pressed

MIX TOGETHER all ingredients (except tofu) in a bowl.

Cut block of tofu lengthwise once, and then 10 times widthwise, to make 20 squares.

Gently toss the tofu in the marinade and transfer to an airtight container with a tight-fitting lid, such as a mason jar.

Let marinate overnight.

Store in the fridge, but bring to room temperature before using.

YIELD: 2 cups (425 g)

BAKED MACADAMIA NUT BRIE

 Cheese sub: silken tofu for milk, macadamia nuts for creaminess

 Corn Free **Wheat Free**

Other than the baking time, this is a brie-ze to throw together.

- 14 ounces (397 g) silken tofu
- 3 ounces (85 g) dry-roasted macadamia nuts
- 2 tablespoons (16 g) arrowroot powder
- 2 tablespoons (42 g) agave nectar
- 1 tablespoon (18 g) white or yellow miso
- 1 tablespoon (15 ml) apple cider vinegar
- 1 teaspoon ground mustard seed
- 1/2 teaspoon sea salt
- Nonstick cooking spray

PREHEAT OVEN to 350°F (180°C, or gas mark 4).

Add all ingredients to a blender and blend until silky smooth.

Coat a small round casserole (about 6 inches [15 cm]) in diameter or 4 small ramekins with cooking spray.

Pour mixture into baking dish, making sure to leave at least a half inch (1.3 cm) at the top to allow for rising.

Place baking dish(es) on a baking sheet (in case it cooks over) and bake for 30 minutes for small ramekins, or 45 minutes for small casserole.

Remove from oven when surface is slightly firm to the touch and starting to brown.

Place in the fridge and allow to cool completely before serving.

To serve, invert on plate and top with fresh berriers or fruit compote, if desired.

YIELD: 1 large or 4 small wheels of brie

NUTTY PEPPERJACK

 Cheese sub: raw cashews for creaminess, nutritional yeast for cheesy flavor

 Corn Free **Gluten Free**

 Soy Free **Wheat Free**

With just a touch of heat from the peppers, this Jack-like loaf tastes great sliced on crackers and in sandwiches, and is especially wonderful in quesadillas.

- Nonstick cooking spray
- 1 ounce (28 g) agar flakes or powder
- 3 cups (705 ml) water
- 2 cups (275 g) raw cashews, finely ground into a powder
- 3 tablespoons (45 ml) fresh lemon juice
- 2 tablespoons (30 ml) olive oil
- 1/4 cup (30 g) nutritional yeast
- 2 teaspoons fine sea salt
- 1/2 teaspoon onion powder
- 1/2 teaspoon garlic powder
- 1/4 cup (25 g) jarred sliced jalapeños, diced
- 1/4 cup (25 g) diced pimiento peppers (or roasted red peppers)

PREPARE A 9 x 5-inch (23 x 13 cm) loaf pan (or other deep rectangular container) by lightly oiling or coating with cooking spray.

Place agar in water and bring to a full boil. Boil for 5 minutes, whisking regularly.

Place all other ingredients into a food processor and blend until smooth.

Pour into the agar mixture and mix until creamy and smooth.

Stir in the peppers. Remove from heat.

Quickly pour into oiled loaf pan and refrigerate until hardened.

YIELD: 1 loaf

Clockwise from top: Marinated Buffalo-style
Tofu Mozzarella in jar (recipe on page 47),
Baked Macademia Nut Brie, Nutty Pepperjack

I'VE GOT THE BLUES DRESSING

 Cheese sub: nondairy sour cream for sour cream, vegan mayo for mayo, and extra-firm tofu for bleu cheese crumbles

 Quick and Easy

 Wheat Free

Why the blues? Because no one, not even us, has come up with a reputable vegan alternative to moldy blue cheese. Even so, this dressing is a bit tangy, has a nice creamy texture, and thanks to the crumbled tofu, the mouthfeel, if not the exact flavor, of blue cheese dressing.

1/4 cup (60 ml) unsweetened soymilk
1 tablespoon (15 ml) lemon juice
1/4 cup (60 g) nondairy sour cream, store-bought or homemade (page 20)
2 tablespoons (28 g) vegan mayonnaise, store-bought or homemade (page 66)
1 tablespoon (4 g) chia seeds
1 tablespoon (15 ml) white wine vinegar
1 teaspoon agave nectar
1/4 teaspoon garlic powder
2 ounces (56 g) extra-firm tofu, crumbled
Sea salt and freshly ground black pepper to taste

ADD LEMON JUICE to milk and let curdle so it resembles buttermilk. Whisk together remaining ingredients and store in an airtight container in the refrigerator until ready to use.

YIELD: 1 cup (235 ml)

MAC & CHEESE MIX

 Cheese sub: cashews for creaminess, nutritional yeast for cheesy flavor

 Corn Free   Gluten Free

 Quick and Easy   Soy Free

 Wheat Free

As easy as ripping open a packet of orange powder, this mix will make a quick task of making kid-friendly mac and cheese.

3 cups (413 g) raw cashews
2 cups (240 g) nutritional yeast
1/2 cup (64 g) arrowroot powder
3 tablespoons (24 g) garlic powder
3 tablespoons (24 g) onion powder
1 tablespoon (18 g) sea salt
1 tablespoon (8 g) ground mustard seed
2 teaspoons paprika
1 teaspoon dried parsley
1 teaspoon dried green onion
1/2 teaspoon turmeric
1/2 teaspoon ground black pepper
1/4 teaspoon cayenne pepper (or chili powder for less heat)
1/4 teaspoon ground cumin

USING A VERY DRY BLENDER or a coffee grinder, grind cashews in small batches into a very fine powder.

In a container with a tight-fitting lid, add all other ingredients and shake vigorously until well mixed.

Refrigerate for up to a month, or freeze.

To use, combine a heaping 1/2 cup (about 95 g) mix with 1 cup (235 ml) water (or nondairy milk for extra creaminess) in a sauce pot. Stir over medium heat until thickened and add to 1 pound (454 g) of prepared pasta.

YIELD: 5 cups (950 g), enough for 10 batches of prepared mac and cheese

CHEESY NO-COOK SAUCE

 Cheese sub: nondairy milk for creaminess, nutritional yeast and maca powder for cheesy flavor

 Quick and Easy Wheat Free

This creamy sauce takes no time to prepare, thanks to the magic of xanthan gum, and it can be tailored to suit your taste as it is extremely versatile in how you season it.

- 1 cup (235 ml) unsweetened nondairy milk
- 1/4 cup (30 g) nutritional yeast
- 1/2 teaspoon fine sea salt
- 1/4 teaspoon ground white pepper to taste
- 1 1/2 teaspoons whole-grain mustard
- 1 teaspoon granulated onion
- 1 large clove garlic, peeled and chopped
- 2 tablespoons (30 ml) extra-virgin olive oil
- 1 tablespoon (15 g) maca powder, optional but recommended
- 1/2 teaspoon xanthan gum

USING A BLENDER, combine all ingredients until thickened. This should take up to 2 minutes.
   Chill the sauce if you want it to thicken even more. Serve hot, on top of gratins, or as a dip for veggies.

YIELD: 1 1/4 cups (265 ml)

WALNUT "PARMESAN" SPRINKLES

 Cheese sub: walnuts for Parmesan, nutritional yeast for cheesy flavor, panko for texture

 Quick and Easy

Make these and store them in a shaker jar to add flavor to pasta, pizza, salads, or anywhere you'd want to sprinkle some grated Parmesan.

- 1/2 cup (60 g) walnut pieces
- 1/2 cup (60 g) nutritional yeast
- 1/2 cup (40 g) panko bread crumbs
- 1/2 teaspoon dried basil
- 1/4 to 1/2 teaspoon salt to taste

ADD ALL INGREDIENTS to a blender or food processor and pulse to combine until walnut pieces are ground into a powder. Store in an airtight container in the fridge. Keeps for weeks.

YIELD: 1 1/2 cups (126 g)

NACHO QUESO

 Cheese sub: soy creamer for heavy cream, nutritional yeast for cheesy flavor, cashews for creaminess, and miso for sharp cheddar-like flavor

 Wheat Free

Spicy, creamy, and perfect poured over a pile of tortilla chips!

- 2 cups (470 ml) soy creamer
- 1/2 cup (80 g) nutritional yeast
- 1/2 cup (70 g) raw cashews, ground into a fine powder
- 1 tablespoon (16 g) tahini
- 2 tablespoons (36 g) mellow white miso
- 2 tablespoons (16 g) cornstarch
- 1 tablespoons (8 g) onion powder
- 1 tablespoon (8 g) garlic powder
- 1 tablespoon (8 g) ground mustard
- 1 teaspoon cumin
- 1 teaspoon hot sauce, like Tabasco, add more if you like it really hot
- 8 to 10 pieces of jarred sliced jalapeños
- 1 tablespoon (15 ml) juice from the jar of jalapeños

USING A BLENDER, combine all ingredients until very smooth.

Pour into a saucepot and stir constantly over medium heat until mixture thickens to desired consistency.

YIELD: 2¹/₂ cups (588 ml)

SPIRALS WITH "FETA" AND SPINACH

 Cheese sub: tofu feta for feta

How about a lighter but oh-so-satisfying pasta dish?

- 1 pound (454 g) uncooked spiral pasta
  One recipe Tofu Feta (page 46)
- 1 tablespoon (10 g) minced garlic
- 1 shallot, diced
- 2 tablespoons (30 ml) olive oil
- 2 cups (60 g) baby spinach
- 1/2 cup (60 g) pine nuts
- 1/3 cup (80 ml) balsamic vinegar
  Salt and pepper

PREPARE PASTA according to package directions.

Prepare Tofu Feta and set aside.

While pasta is boiling, sauté garlic and shallot in olive oil for 5 minutes, or until fragrant and translucent.

Drain pasta and return to the pot.

Add feta, spinach, and pine nuts. In a small bowl, whisk together vinegar and sautéed mixture.

Add mixture to the pot and toss so all ingredients are well incorporated.

Add salt and pepper to taste.

Serve warm or cold.

YIELD: 4 to 6 servings

FONDUE

 Cheese sub: nutritional yeast, maca powder, and miso for cheesy flavor and sharpness

 Corn Free   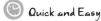 Quick and Easy

This cookbook could not be co-written by a Swiss person if it did not include a recipe for fondue! This delicious concoction will be thin when cold but will thicken back up once heated.

- 2 cups (470 ml) dry white wine
- 1¹/2 cups (455 ml) unsweetened nondairy milk
- ¹/2 cup (60 g) nutritional yeast
- 2 tablespoons (30 g) maca powder, optional but strongly recommended
- 3 tablespoons (48 g) tahini
- 1 tablespoon (18 g) reduced-sodium white miso
- ¹/4 cup (40 g) potato starch
- 2 tablespoons (30 ml) white wine vinegar
- 1 tablespoon (7 g) granulated onion
- 1 clove garlic, peeled and chopped
- 1 teaspoon whole-grain mustard
- ¹/2 teaspoon ground white pepper to taste
- Pinch sea salt to taste
- 1¹/2 to 2 teaspoons Kirsch, optional (makes for a more authentic taste)

BRING WINE to a boil in a large saucepan. Lower heat to a simmer. Use a blender to combine the rest of the ingredients until smooth.

Pour blended mixture into wine, whisking immediately to avoid lumps. Cook until just thickened, about 2 minutes.

Dip bread cubes or veggies into fondue. If you have leftovers and need to reheat the fondue, you might have to add extra nondairy milk to thin it out a bit.

YIELD: 3¹/2 cups (825 g)

Variation: Make Croûtes au Fromage by preheating oven to 375°F (190°C, or gas mark 5). Add 2 tablespoons (42 g) or more fondue and a slice of tomato to each thick slice of crusty bread, adding a sprinkle of olive or truffle oil on top, and baking for 15 minutes or until the fondue is browning up and bubbling on the bread.

INDIAN-SPICED PUMPKIN GRATIN

 Cheese sub: coconut milk for richness, pumpkin with turmeric for color

Who needs cheese when spices and coconut naturally impart delicious flavors to meals such as this beautifully orange-hued gratin that gets better every day?

FOR FILLING:
- 2 medium potatoes, chopped
- 1 pound (454 g) uncooked macaroni
- 2 teaspoons cumin seeds
- 1 teaspoon mustard seeds
- 2 tablespoons (28 g) coconut oil, melted
- 1 can (15 ounces, or 425 g) chickpeas, drained and rinsed
- 2 cloves garlic, peeled and minced
- 1 teaspoon ground coriander
- 6 cups (180 g) baby spinach

FOR PUMPKIN SAUCE:
- 1 cup (244 g) pumpkin purée
- 2 tablespoons (24 g) Sucanat
- 2 tablespoons (30 g) tamarind paste
- 1/2 teaspoon red pepper flakes
  Pepper to taste
- 1/2 teaspoon sea salt

- 1 teaspoon turmeric
- 1/2 teaspoon ground cinnamon
- 1/2 teaspoon ground cardamom
- 2 teaspoons granulated onion
- 1 cup (235 ml) heated vegetable broth

FOR COCONUT SAUCE:
- 1 cup (235 ml) coconut milk
- 1/2 teaspoon curry powder
- 1/4 teaspoon ginger powder
  Pepper to taste
- 1/2 teaspoon sea salt
- 1/2 teaspoon red pepper flakes

TO MAKE THE FILLING: Place potatoes in a large pot filled with 5 quarts (4.7 L) water. Bring to a boil, cook for 4 minutes. Add pasta and cook according to package instructions. Drain and set aside.

Toast seeds in oil in medium saucepan. Add chickpeas, garlic, and coriander. Let brown for 2 minutes over medium heat. Coarsely mash chickpeas.

Add spinach, and cook until wilted, about 2 minutes. Set aside.

TO MAKE THE PUMPKIN SAUCE: Combine all ingredients with broth. Combine with pasta and chickpeas.

TO MAKE THE COCONUT SAUCE: In a medium bowl, whisk all ingredients except for pepper flakes.

Preheat oven to 375°F (190°C, or gas mark 5). Prepare a 9-inch (23 cm) square, deep baking dish. Press down pasta into dish. Pour coconut sauce on pasta. Sprinkle with red pepper flakes. Place dish on a rimmed baking sheet.

Bake for 20 minutes, or until light golden brown. Let stand for 15 minutes before serving.

YIELD: 6 servings

SAVORY ARTICHOKE PIE

 Cheese sub: nutritional yeast for cheesy flavor

With a flaky and cheesy crust that is a snap to make, this flavorful pie offers great textures and creaminess.

FOR CRUST:

Nonstick cooking spray
1¹/₂ cups (188 g) all-purpose flour
¹/₄ cup (30 g) nutritional yeast
1 clove garlic, peeled and chopped
¹/₂ teaspoon ground black pepper
1 teaspoon fine sea salt
¹/₄ cup plus 1 tablespoon (70 g) nondairy butter
¹/₄ cup (60 ml) water

FOR FILLING:

1 can (14 ounces, or 397 g) artichoke hearts, chopped
2 teaspoons peanut oil
Pinch sea salt
Pinch red pepper flakes
Pinch ground nutmeg
1 cup (270 g) Roasted Squash and Lentil Spread (page 256)

FOR TOPPING:

1 can (15 ounces, or 425 g) chickpeas, drained and rinsed
¹/₄ cup (30 g) nutritional yeast
1 clove garlic, peeled and chopped
Zest of a lemon
¹/₂ cup (112 g) reduced-fat vegan mayonnaise
¹/₂ cup (120 ml) unsweetened nondairy milk
¹/₂ teaspoon ground black pepper
¹/₂ teaspoon sea salt

PREHEAT OVEN to 375°F (190°C, or gas mark 5). Lightly coat a 9-inch (23-cm) round baking pan with cooking spray.

TO MAKE THE CRUST: Combine flour, yeast, garlic, pepper, and salt in food processor. Pulse to combine.

Add butter. Pulse to reach consistency of a coarse meal. Add water, 1 tablespoon (15 ml) at a time, until dough holds together when pinched.

Crumble dough evenly into pan. Press down at the bottom and edges.

Prebake for 18 minutes, until edges turn golden brown. Set aside.

TO MAKE THE FILLING: Combine artichoke hearts, oil, salt, and spices in medium saucepan.

Cook over medium heat until artichokes brown, about 6 minutes. Add Roasted Squash and Lentil Spread; stir to combine.

Place filling evenly into crust.

TO MAKE THE TOPPING: Combine all ingredients in blender or food processor. Blend until smooth. Spread evenly on top of filling.

Bake for 20 minutes, or until golden brown. Let stand for 15 minutes before serving.

YIELD: 8 slices

BAKED PESTOTATOES

Cheese sub: firm tofu for creaminess

This fresh and creamy pesto yields enough deliciousness to fill about 12 baked potatoes. The pesto can also be used as a sandwich spread, dip for crudités, or served with pasta.

FOR PESTO:

- 19 ounces (539 g) firm tofu
- 1 1/2 cups (60 g) fresh basil leaves
- 1/4 cup (60 ml) apple cider vinegar
- 1/4 cup (60 ml) extra-virgin olive oil
- 3/4 teaspoon fine sea salt
  White pepper to taste
- 1/2 teaspoon red pepper flakes to taste
- 2 large cloves garlic, peeled and chopped
- 2 teaspoons granulated onion
- 1 heaping teaspoon whole-grain Dijon mustard
- 1 1/2 teaspoons agave nectar
- 2 tablespoons (30 ml) fresh lemon juice
- 1 teaspoon fresh thyme

FOR BAKED POTATOES:

- 2 large russet potatoes, brushed clean, unpeeled, lightly rubbed with peanut oil, pierced with a fork about 10 times

TO MAKE THE PESTO: Use an immersion blender or a food processor. Blend all ingredients until smooth. Store in the fridge.

TO MAKE THE BAKED POTATOES: Preheat oven to 375°F (190°C, or gas mark 5).

Place potatoes directly on upper rack, with a baking pan on lower rack to catch drippings. Bake for 1 hour, or until tender. Let sit until cool enough to handle. Carefully hollow out the potato by cutting out a hole in the center of the potato (keep it in one piece—you will place it back after adding the pesto!). Place 2 heaping tablespoons (60 g) pesto in each potato. Place hat back on top of potato, bake for another 10 minutes.

Serve immediately.

YIELD: 3 cups (780 g) pesto; 2 pesto-filled potatoes

CREAMY POLENTA CHILI BAKE

 Cheese sub: nutritional yeast for cheesy flavor

This chili comes topped with a deliciously creamy polenta sauce, which is sure to rival and no doubt surpass any cheese sauce you've tasted before.

FOR CHILI:
- 1 can (6 ounces, or 170 g) tomato paste
- 1/4 cup (40 g) chopped onion
- 2 cloves garlic, peeled and minced
- 1 teaspoon ground cumin
- 1 teaspoon Mexican oregano
- 1 tablespoon (8 g) mild to medium chili powder
- 1/2 cup (120 ml) dry white wine
- 1 teaspoon vegan Worcestershire sauce
- 1 can (15 ounces, or 425 g) kidney beans, drained and rinsed
- 1 can (15 ounces, or 425 g) black beans, drained and rinsed
- Water, optional, as needed, to bring chili to desired consistency

FOR POLENTA:
- 2 cups (470 ml) unsweetened nondairy milk
- 1/4 cup (30 g) nutritional yeast
- 1 teaspoon fine sea salt
- 1/2 teaspoon black pepper
- 1/2 cup (70 g) cornmeal
- 1 tablespoon (14 g) nondairy butter
- Cayenne pepper to taste, optional

TO MAKE THE CHILI: Combine the first 8 ingredients in a small saucepan. Bring to a low boil. Lower heat, and add beans. Add optional water, as needed, to reach desired consistency. Simmer for 10 minutes. Set aside.

TO MAKE THE POLENTA: Combine the first 4 ingredients in a small saucepan.
Bring to a low boil.
Slowly whisk in cornmeal to avoid lumps. Cook for 8 minutes, stirring often. Stir in butter until melted. Set aside.
Preheat oven to 375°F (190°C, or gas mark 5). Place chili in a nonstick 9 x 5-inch (23 x 13 cm) loaf pan. Top evenly with polenta, and sprinkle cayenne pepper on top if desired. Bake for 30 minutes.
Let cool for 10 minutes before serving.

YIELD: 3 servings

MAKE THE CHICKENS SMILE!

FOOLPROOF SUBSTITUTIONS FOR EGGS

Chapter 3
From Tender Cakes to Scrumptious Scrambles:
HOW TO SUBSTITUTE FOR EGGS

ONE OF THE MOST COMMON QUESTIONS asked of vegans is, "How do you bake without eggs?"

Easily, in fact! Contrary to popular belief, eggs haven't always been readily available, and vegans weren't the first ones to have to come up with ways to replace them. Consider the famous Wacky Cake, created ages ago due to wartime rationing, based on the simple combination of vinegar and baking soda. It is to this day the most famous and beloved, "accidentally vegan" recipe available.

For the most part, eggs are used in recipes as emulsifying, binding, leavening, or structure-giving agents. Fortunately, there are many vegan-friendly ingredients that lend themselves quite nicely to these exact same purposes.

Consider the Facts

Many people don't see the harm in eating eggs. After all, you needn't kill the hen to get her eggs, right? While technically this may be true, the living conditions of even so-called cage-free chickens are horrendous.

Consider the fact that a mere 30 years ago, hen-laying chickens weighed an average of 2 to 3 pounds and laid approximately 50 to 60 eggs per year. Today, a chicken typically weighs about 5 to 6 pounds and lays about 175 or more eggs per year. If that's not genetic engineering, what is? (For more information on the horrors of the egg industry, visit NoEggs.com.)

BAD FOR THE CHICKENS AND BAD FOR YOU, TOO

The egg industry would have you believe that eating eggs is both harmless to chickens and rather healthy for you. In truth, however, eggs rank toward the bottom of the health meter: They have been shown to contribute to allergies, weight problems, and other health issues, even early death.

How could that be? One Grade A large (50 g) egg contains 71 calories, 5 grams of fat (2 g saturated fat), 70 mg of sodium, 6 grams of protein, and a whopping 211 mg of cholesterol. As one gram of fat is equal to nine calories, we can see that this artery-clogging "food" gets more than half of its calories from fat. Certainly doesn't sound like a health food to us!

Egg Substitution Guidelines

Now that you know why eggs aren't so golden, let's focus our attention on how we can substitute for them in our cooking.

When determining the function of the egg you are looking to replace, consider the following:

- If you're making an airy and fluffy cake or bread, you want the egg substitute to act as a leavening agent. You can usually tell the egg is needed as a leavening agent if the recipe you are working from requires 2 to 3 eggs.

- For chewy and sometimes crispy cookies, your substitute will function more as a moisturizing and binding agent. You can usually tell the egg is needed as a binding agent if the recipe you are working from requires 1 egg.

- When making savory dishes, such as gnocchi or meatloaf, the eggs generally act as binders.

- When it comes to quiches and scrambles, where eggs are the star of the show, tofu is your go-to substitution.

For more specific guidelines, let us turn to the following chart, and then see if we can apply our knowledge to a nonvegan recipe.

IF THE ORIGINAL RECIPE CALLS FOR...	REPLACE WITH...
1 egg in baked goods, for binding; usually 1 egg per recipe means it is used for binding (e.g., cookies and cakes)	• 1½ teaspoons Ener-G or Bob's Red Mill egg replacer powder whisked with 2 tablespoons (30 ml) warm water • 2 tablespoons (16 g) cornstarch or arrowroot, or any starch whisked with 2 tablespoons (30 ml) water • 2½ tablespoons (18 g) flaxseed meal whisked with 3 tablespoons (45 ml) warm water • 1/4 cup (60 g) blended silken tofu • 1/4 cup (60 g) applesauce, pumpkin, or other fruit or vegetable purée
1 egg in baked goods, for leavening; usually 2 to 3 eggs per recipe means they are used for leavening (e.g., fluffy cakes, muffins, or quick bread)	• 1½ teaspoons Ener-G or Bob's Red Mill egg replacer powder whisked with 2 tablespoons (30 ml) warm water • 1 tablespoon (15 ml) mild-flavored vinegar combined with nondairy milk (soymilk yields best results) to curdle and make 1 cup (235 ml)—works best used in recipes that involve baking soda • 1/4 cup (60 g) nondairy yogurt

IF THE ORIGINAL RECIPE CALLS FOR...	REPLACE WITH...
1 egg in baked goods, for moisture (e.g., cakes and cookies)	• $^1/_4$ cup (60 ml) coconut milk • 1 teaspoon oil or nut/seed butter combined with nondairy milks, to make $^1/_4$ cup (60 ml) • $^1/_4$ cup (60 g) fruit or vegetable purée
1 egg white	• $1^1/_2$ teaspoons Ener-G or Bob's Red Mill egg replacer powder whisked with 2 tablespoons (30 ml) warm water • Dissolve 1 tablespoon (8 g) agar powder in 1 tablespoon (15 ml) water, whip, chill thoroughly, and whip again. (We do not recommend this for recipes that call for more than 2 egg whites, as it becomes more difficult to substitute for a larger quantity of eggs.)
1 egg in savory foods as a binder (e.g., casseroles, meatless loaf, and batters for frying)	• $1^1/_2$ teaspoons Ener-G or Bob's Red Mill egg replacer powder whisked with 2 tablespoons (30 ml) warm water • $^1/_4$ cup (60 g) blended silken tofu • $2^1/_2$ tablespoons (18 g) flaxseed meal whisked with 3 tablespoons (45 ml) warm water • 2 tablespoons (33 g) tomato paste or vegetable purée or unsweetened nut/seed butters, as a moisturizing and binding agent, as in Meatfree Balls (page 114) • 2 tablespoons (15 g) flour, starches, bread crumbs, as a binding agent, quantity as needed. Start by adding moisture to your meatless loaf or quiche, then add binding agent until batter holds together, as in Meatfree Balls (page 114) or Denver Quiche (page 70)
1 egg in savory foods as a leavening agent for rise and structure (e.g., soufflés)	• $^1/_4$ cup (60 g) blended silken tofu
Scrambled eggs	• Scrambled tofu as in Spinach and Mushroom Tofu Scramble (page 72)
Hard-boiled eggs	• Extra- or super-firm tofu as in Traditional "Egg" Salad (page 68)

VEGANIZED!: SAMPLE RECIPE

Let's look at a traditional recipe for an example of how we would replace the eggs and other nonvegan ingredients:

SNICKERDOODLES

This cinnamon and sugar cookie has been adapted from the *Better Homes and Gardens New Cook Book*.

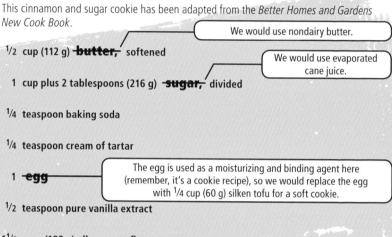

> We would use nondairy butter.

1/2 cup (112 g) ~~butter,~~ softened

> We would use evaporated cane juice.

1 cup plus 2 tablespoons (216 g) ~~sugar,~~ divided

1/4 teaspoon baking soda

1/4 teaspoon cream of tartar

1 ~~egg~~

> The egg is used as a moisturizing and binding agent here (remember, it's a cookie recipe), so we would replace the egg with 1/4 cup (60 g) silken tofu for a soft cookie.

1/2 teaspoon pure vanilla extract

1 1/2 cups (188 g) all-purpose flour

1 teaspoon ground cinnamon

BEAT NONDAIRY BUTTER with an electric mixer on medium to high speed for 30 seconds.

Add 1 cup (192 g) evaporated cane juice, baking soda, and cream of tartar. Beat until combined, scraping sides occasionally.

Add blended tofu and vanilla, and beat until well combined. Beat in flour, using the mixer. Cover and chill dough for 1 hour.

Preheat oven to 375°F (190°C, or gas mark 5).

Line 2 cookie sheets with parchment paper or silicone baking mats.

Combine remaining 2 tablespoons (24 g) evaporated cane juice with cinnamon. Set aside.

Shape dough into 1-inch (2.5 cm) balls. Roll balls in cinnamon sugar mixture to coat.

Place 2 inches (5 cm) apart on prepared cookie sheets.

Bake for 10 to 12 minutes, or until edges turn golden. As soon as the cookies are firm enough, transfer from cookie sheets to a wire rack and let cool completely before storing.

YIELD: 36 cookies

Finding Egg Substitutes at the Store

There are several commercially available egg replacers on the market. These products make an easy task of turning virtually any recipe into an egg-free delight! Two of the brands we like are Ener-G and Bob's Red Mill. Personal experience has taught us that Ener-G tends to work more favorably in savory dishes, while Bob's Red Mill lends itself well to baking. Both are available at health-food stores and online. The usual rule for using these products is to combine by whisking 1 1/2 teaspoons egg replacer powder with 2 tablespoons (30 ml) warm water.

When it comes to another egg-based product, mayonnaise, we like Vegenaise as it has the best flavor and is now available in a low-fat version.

Making Egg Substitutes at Home

Not a fan of commercial egg replacers? Here are several ways to replace eggs using ingredients you probably already have in the kitchen. In the following pages, you will find lots of recipes utilizing these handy techniques.

USE AGAR POWDER

If you are looking to replace just the white part of the egg in your baking recipe, dissolve 1 tablespoon (8 g) agar powder in 1 tablespoon (15 ml) water, whip, chill thoroughly, and whip some more.

USE APPLESAUCE

Simply replace 1/4 cup (60 g) unsweetened applesauce (store-bought or homemade, page 231) for each egg called for in a recipe. This works best in baking recipes, such as cakes, soft cookies, or muffins. The flavor won't overwhelm the recipe. Adding an extra 1/2 teaspoon baking powder will help obtain a lighter texture in baked goods.

USE BANANAS

Mash 1/2 a ripe banana to get about 1/4 cup (60 g) purée to replace each egg called for in a recipe. This works best in baking recipes, such as cakes, cookies, or muffins. Be careful if you are not a fan of the banana flavor, as it will often take the lead in the recipe.

USE BINDING AGENTS IN SAVORY ITEMS

When making veggie burgers and meatless balls, flours and starches, oats, potato flakes, breadcrumbs, and even tomato paste and applesauce do a great job at replacing the egg in such recipes. Use 2 tablespoons (15 g for starches and flours, 33 g for tomato paste) to replace 1 egg.

USE COCONUT MILK

Use $1/4$ cup (60 ml) coconut milk per egg. The amount of fat naturally contained in coconut milk replaces the fat found in eggs. Be careful if you are not a fan of the coconut flavor, as it tends to dominate the recipe.

USE DRIED FRUIT

Combine equal parts of any dried fruit and water in saucepan. Simmer for 15 minutes. Blend. Substitute $1/4$ cup (about 71 g, depending on fruit) of the mixture per egg called for in a recipe. We're big fans of using dried plums or dates, but we haven't met a dried fruit that didn't yield great results.

USE FLAX SEED

Whisk $2^1/2$ tablespoons (18 g) ground flaxseeds with 3 tablespoons (45 ml) warm water to replace one beaten egg called for in a recipe. The mixture will become gelatinous. Keep in mind that this mixture will lend a nutty flavor to your recipe.

USE FLOURS AND STARCHES

2 tablespoons (30 ml) water whisked with 2 tablespoons (approximately 15 g) flour, such as all-purpose, or starch, such as potato, tapioca, arrowroot, or corn, can be used in a similar fashion as store-bought powdered egg replacer to replace one egg. Best if used in recipes such as puddings, casseroles, and other egg-heavy dishes. See Chocolate Pudding (page 36) and Amaretti Cookies (page 189) for examples of use.

USE HOMEMADE MILK PULP

Don't throw away the pulp that is left behind once you're done preparing homemade milks! Add $1/4$ cup (45 g) leftover pulp blended into the liquid ingredients called for in the recipe to replace one beaten egg. See Pulp Not Fiction Muffins (page 75) for example of use.

USE NONDAIRY PUDDING

Replace $1/4$ cup (60 g) pudding (store-bought or homemade [page 36]) for each egg needed in a recipe. This technique works well in baking. You can adjust the flavor of pudding used to suit the needs of the recipe.

USE NONDAIRY YOGURTS, SOUR CREAM, OR MAYO

Replace ¼ cup (60 g) nondairy yogurt (store-bought or homemade [page 20]) for each egg needed in a recipe. This technique works well in sweet and savory recipes. You can adjust the flavor of yogurt used to suit the needs of the recipe. Consider using ¼ cup (60 g) nondairy sour cream or mayonnaise if you run out of yogurt: It won't change the flavor but contains more fat than yogurt and is often harder to find and a bit more costly.

USE NUT BUTTERS

Mix 1 teaspoon of nut butter (store-bought or homemade [page 251]) into ¼ cup (60 ml) nondairy milk. Works best in cookie recipes. Be careful if you are not a fan of strong-flavored nut butters, such as peanut, as it will take the lead in the recipe.

USE OIL

Mix 1 teaspoon of any neutral-flavored oil into ¼ cup (60 ml) nondairy milk. Works best in cookie recipes.

USE TOFU

Add ¼ cup (60 g) silken tofu, blended into the liquid ingredients called for in the recipe to replace one beaten egg. Works best in muffins and cakes, as tofu contributes to a softer texture.

USE VEGETABLE PURÉE

Pumpkin, carrot, butternut squash, and sweet potato purées work especially well. Add ¼ cup (60 g) to replace one egg. Works well in cake or muffin recipes.

QUICK TIP!

Remember that strong-flavored egg replacers, such as bananas or peanut butter, will affect the taste of the final dish, so avoid using them in recipes where they may not complement the other ingredients.

The Recipes: Egg-Based Recipes... without the Eggs!

Start with the basics and work your way toward the more difficult. From mayonnaise, muffins, cakes, and cookies, to scrambles and quiches—you'll be whipping up some great dishes without ever having to crack an egg!

· ·

BASIC TOFU MAYO

 Corn Free   Quick and Easy   Wheat Free

This recipe works very well as a sandwich spread or in any of the mayo-based dressings in this book. And as long as you use gluten-free vinegar, this recipe is gluten free, too.

   7 ounces (198 g) extra-firm tofu, drained and pressed
   1/4 cup (35 g) raw cashews, ground into a very fine powder
   1 tablespoon (15 ml) lemon juice
   1 tablespoon (12 g) raw sugar or (21 g) agave nectar
   1 1/2 teaspoons Dijon mustard
   1 teaspoon apple cider vinegar
   1/2 teaspoon sea salt
   1/4 cup plus 2 tablespoons (90 ml) canola oil
     or other mild-flavored vegetable oil

PLACE ALL INGREDIENTS except oil in a blender or food processor and blend until smooth.
   Slowly drizzle in the oil until you get the consistency that you like.
   Store refrigerated in an airtight container for up to 2 weeks.

YIELD: 2 cups (470 ml)

BASIC SOYMILK MAYO

 Nut Free   Wheat Free

Tired of paying for mayonnaise at the store? Try this simple, tasty version that also makes for a great gratin topping. Get creative and switch seasonings depending on use and preference!

1 cup (235 ml) unsweetened soymilk
1 to 2 tablespoons (15 to 30 ml) apple cider vinegar or fresh lemon juice or a combination of both
1^1/$_2$ teaspoons prepared mustard of choice
^{1}/$_4$ teaspoon garlic powder
^{1}/$_2$ teaspoon dried minced onion
Generous pinch dried dill
^{1}/$_2$ teaspoon fine sea salt
^{1}/$_4$ teaspoon ground pepper
1^1/$_2$ tablespoons (12 g) cornstarch
^{1}/$_4$ cup (60 ml) extra-virgin olive oil

COMBINE SOYMILK and vinegar in a medium bowl; let curdle for a few seconds.
    Add remaining ingredients. Using an immersion blender, blend for 1 minute.
    Heat in microwave for 1 minute; watch to make sure the mixture doesn't bubble up.
Blend again.
    Heat in microwave for 1 minute; blend again.
    Heat 1 last minute; let stand to cool.
    Once cooled, stir with a fork.
    Transfer to an airtight container. Chill in the fridge to thicken. Stir with a fork before using.
Keep stored in the fridge for up to a week.

YIELD: 1 generous cup (250 g)

Note: Make it into a no-cook mayo: Replace cornstarch with 1 scant teaspoon xanthan gum, and blend thoroughly until thickened. The mayo will thicken even more in the fridge. This will work with any nondairy milk.

"HARD-BOILED EGG" SALAD

 Egg sub: nutritional yeast and maca powder

 Corn Free   Nut Free

 Quick and Easy   Soy Free

We used to enjoy hard-boiled egg salad as kids, and the flavor of this recipe reminds us of it, thanks to the combination of nutritional yeast and maca powder, which also happens to be a great source of calcium.

1 tablespoon (10 g) chopped onion
1 tablespoon (7 g) nutritional yeast
1 teaspoon maca powder
1 tablespoon (15 ml) extra-virgin olive oil
1 tablespoon (15 ml) white balsamic vinegar
1 medium tomato, chopped
  White pepper to taste
  Pinch black salt to taste, optional
2 cups (75 g) packed fresh baby spinach

COMBINE ALL INGREDIENTS except spinach in a large bowl.
  Once combined, add spinach and mix together.

YIELD: 1 main-dish or 2 side-dish servings

TRADITIONAL "EGG" SALAD

 Egg sub: tofu for hard boiled eggs

 Corn Free   Nut Free

 Quick and Easy   Wheat Free

Without any of the add-ins, this salad's texture may be a little runny; to thicken, don't use all of the dressing.

FOR DRESSING:
14 ounces (397 g) silken tofu, drained but not pressed
1/2 cup (120 ml) canola oil
2 tablespoons (30 ml) yellow mustard
1 tablespoon (15 g) minced garlic
1 tablespoon (18 g) mellow white or yellow miso
1/2 teaspoon black salt

FOR EGGS:
20 ounces (567 g) extra- or super-firm tofu, drained and pressed
1 teaspoon turmeric
  Salt and pepper to taste

OPTIONAL ADD-INS:
  Chopped celery, pickle relish, chopped fresh parsley, chopped fresh green onions

TO MAKE THE DRESSING: Purée all ingredients in a blender until smooth.

TO MAKE THE EGGS: Mash half of the tofu in a mixing bowl with the turmeric until it resembles crumbled egg yolks and the turmeric is well incorporated. This is the "egg yolks." Cut the remaining tofu into tiny cubes. This will be the "egg whites." Combine the whites and yolks and mix in the dressing. Add optional add-ins, with salt and pepper to taste.

YIELD: 6 cups (800 g), depending on add-ins

EGGS BENEDICT WITH HAM

 Egg sub: tofu for eggs, black salt for sulphury egglike flavor

This diner classic is sure to please! Make the ham ahead of time to save time in the morning.

4 English muffins, halved and toasted

FOR HAM:

- **1/2 cup (60 g) garbanzo or fava bean flour**
- **1/2 cup (72 g) vital wheat gluten**
- **1/2 cup (120 ml) water or vegetable broth**
- **1/4 cup (60 ml) vegetable oil**
- **2 tablespoons (18 g) brown sugar**
- **2 tablespoons (15 g) nutritional yeast**
- **1 1/2 tablespoons (24 g) tomato paste**
- **1/2 to 1 teaspoon liquid smoke to taste**
- **1/2 teaspoon black pepper**
- **1/4 teaspoon salt**
- **Aluminum foil**

FOR FRIED EGGS:

- **2 to 4 tablespoons (30 to 60 ml) vegetable oil**
- **1/4 teaspoon turmeric**
- **1/4 teaspoon black salt** 🐣
- **1/8 teaspoon ground black pepper**
- **1 pound (454 g) extra firm tofu, drained and pressed, cut into 8 thin, squares** 🐣

FOR HOLLANDAISE:

- **1/4 cup (56 g) nondairy butter**
- **2 tablespoons (15 g) all-purpose flour**
- **1 cup (235 ml) plain soy creamer, more if needed**
- **1 tablespoon (15 ml) lemon juice**
- **1 tablespoon (8 g) nutritional yeast**
- **1/8 teaspoon cayenne pepper**
- **Salt to taste**

TO MAKE THE HAM: Preheat oven to 350°F (180°C, or gas mark 4).

Combine all ingredients until goopy mixture forms.

Spread large sheet of foil on counter. Place mixture in the middle. Roll foil over mixture; twist the ends tightly to form a log.

Place foil on a cookie sheet, seam side down; bake for 45 minutes or until firm.

Let cool before unwrapping.

Store in an airtight container. Refrigerate for up to a week or freeze indefinitely.

TO MAKE THE FRIED EGGS: Heat oil in a skillet over medium-high heat.

Combine turmeric, black salt, and pepper.

Add tofu squares to oil.

Sprinkle half of the spice mixture over tofu. Flip after 5 minutes.

Sprinkle remaining spice mixture on the other side of tofu steaks and cook for another 5 minutes.

TO MAKE THE HOLLANDAISE: In a pot, melt butter over high heat.

Add flour and whisk vigorously until smooth. Add creamer, bring to a boil, and remove from heat. Stir in juice, yeast, cayenne, and salt. Add more creamer if the sauce is too thick.

Assemble dish by placing a slice of ham on top of one-half toasted English muffin; place egg on top of ham, and pour generous amount of hollandaise sauce on top. Season to taste.

YIELD: 8 servings

DENVER QUICHE

 Egg sub: tofu for eggs

This eggless quiche can be made crustless or with the flaky crust from the Savory Artichoke Pie (page 55). It's perfect for brunch and tastes great the next day crumbled into a breakfast burrito! The black salt will fool even the pickiest eaters into thinking there's no way this is vegan!

Nonstick cooking spray

14 ounces (397 g) silken tofu, drained but not pressed

1/4 cup (30 g) nutritional yeast

1/4 cup (30 g) chickpea flour

1/4 cup (36 g) cornstarch

1/4 cup (60 ml) olive oil

1 teaspoon onion powder

1 teaspoon garlic powder

1/4 teaspoon turmeric

1 tablespoon (15 g) mild Dijon mustard

1/4 teaspoon black salt

1/2 teaspoon cumin

1/4 teaspoon paprika

1/2 teaspoon liquid smoke

1 cup (160 g) diced onion (red or white)

1 red bell pepper, cored and seeded and diced

1 green bell pepper, cored and seeded and diced

1/4 cup (21 g) imitation bacon bits, store-bought or homemade (page 143)

4 ounces (112 g) diced prepared seitan

PREHEAT OVEN to 350°F (180°C, or gas mark 4).

Prepare a 9-inch (23 cm) pie pan with cooking spray or by inserting a premade or homemade pie crust (page 55).

In a blender, combine tofu, nutritional yeast, chickpea flour, cornstarch, olive oil, onion powder, garlic powder, turmeric, mustard, black salt, cumin, paprika, and liquid smoke.

Purée until smooth.

Transfer mixture to a mixing bowl and stir in onion, red and green bell peppers, bacon bits, and seitan.

Pour mixture evenly into pie pan.

Bake, uncovered, for 45 minutes to 1 hour, or until top is golden brown, and center is firm.

Let cool for at least 10 minutes before slicing.

YIELD: 8 servings

Note: If you prefer a cheesy quiche, fold 1 cup (112 g) shredded vegan cheese into the mixture along with the peppers.

SPINACH AND MUSHROOM TOFU SCRAMBLE

 Egg sub: : tofu for scrambled eggs, black salt for sulphury egg-like flavor

 Corn Free   **Nut Free**   **Quick and Easy**   **Wheat Free**

The tofu scramble is a must in every vegan's repertoire. Once you master the basics, it's a snap to make endless variations of this tasty, protein-packed breakfast staple.

- 2 tablespoons (30 ml) canola or other mild-flavored vegetable oil
- 3/4 cup (113 g) white or yellow onion, chopped
- 1¼ cups (113 g) sliced button or baby bella mushrooms
- 1 tablespoon (15 g) minced garlic
- 2 tablespoons (56 g) jarred, diced pimiento peppers
- 10 to 14 ounces (283 to 397 g) extra-firm tofu, drained and pressed
- 1/4 cup (30 g) nutritional yeast
- 1 teaspoon turmeric
- 2 cups (60 g) baby spinach
- 1 tablespoon (15 ml) mild Dijon mustard
  Salt and pepper to taste
  Pinch black salt, optional

HEAT OIL IN SKILLET over medium-high heat.

Sauté onion and mushrooms for about 5 minutes. Add garlic and pimientos and continue to sauté for 2 more minutes. Crumble in tofu and continue to cook for another minute. Add in yeast and turmeric, and toss to coat. Add in spinach and continue to cook until wilted, about 2 more minutes. Remove from heat and add mustard. Toss to coat.

Add salt and pepper to taste, adding a pinch of black salt if using.

YIELD: 4 servings

SUPER PEANUT PUMPKIN MUFFINS

Egg sub: fruit purée (banana) and vegetable purée (pumpkin) for moisture and binding

There are several egg substitutes in these tasty and nutritious muffins, which will give you superhuman-like strength all morning long! The peanut butter and pumpkin take charge when it comes to flavor, leaving the banana to work on keeping these muffins deliciously tender.

Nonstick cooking spray
1 1/2 cups (180 g) whole wheat pastry flour
2 teaspoons baking powder
1/4 teaspoon fine sea salt
1 teaspoon pure vanilla extract
2 small (2/3 cup, or 160 g) ripe bananas, mashed
1/3 cup (35 g) flaxseed meal
3/4 cup (144 g) cane sugar
1/2 cup (128 g) smooth natural peanut butter
1/2 cup (120 ml) nondairy milk
3/4 cup (183 g) pumpkin purée
1/4 cup (40 g) favorite add-in: chopped nuts, chocolate, or raisins

PREHEAT OVEN to 350°F (180°C, or gas mark 4). Lightly coat a jumbo muffin tin with cooking spray.

Whisk flour, baking powder, and salt in a medium bowl.

In a large bowl, combine vanilla, bananas, flax, sugar, peanut butter, milk, and pumpkin.

Fold dry ingredients into wet, being careful not to overmix.

Stir in add-in of choice.

Divide batter into prepared pan. Bake for 40 minutes, or until a toothpick inserted into center of muffin comes out clean. Remove from the pan and let cool on a wire rack.

YIELD: 6 jumbo muffins

Note: If you don't have a jumbo muffin tin, bake for 25 minutes in a regular muffin pan. You'll get about 12 muffins.

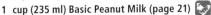

PULP NOT FICTION MUFFINS

 Egg sub: nondairy milk pulp (peanut) for binding

Want to put your homemade milk and its leftover pulp to good use? Enjoy these muffins with a generous amount of our Fruity Chocolate Topper (page 254) or ABC Mousse Topper (page 244).

1¹/2 cups (256 g) Basic Peanut Milk (page 21) pulp
  1 cup (235 ml) Basic Peanut Milk (page 21)
¹/4 cup (60 ml) peanut oil
  2 teaspoons pure vanilla extract
²/3 cup (128 g) Sucanat
¹/2 teaspoon fine sea salt
  2 teaspoons ground cinnamon
¹/2 cup (80 g) raisins or (88 g) semisweet chocolate chips
1¹/2 cups (180 g) whole wheat pastry flour
  1 tablespoon (12 g) baking powder

PREHEAT OVEN to 375°F (190°C, or gas mark 5). Line a standard muffin tin with paper liners.
  Combine pulp, milk, oil, vanilla, Sucanat, salt, and cinnamon in a blender. Blend until smooth.
  Combine raisins or chips, flour, and baking powder in a large bowl.
  Fold wet ingredients into dry, being careful not to overmix.
  Divide batter into prepared muffin tin.
  Bake for 20 minutes, or until a toothpick inserted in the center of a muffin comes out clean.
  Let cool on a wire rack for a few minutes before transferring muffins directly onto rack.

YIELD: 12 muffins

CINNAMON FRENCH TOAST

 Egg sub: just naturally egg free!

 Corn Free   **Nut Free**   **Quick and Easy**

This recipe is so easy it's almost embarrassing. But who doesn't love a ridiculously easy recipe? So here you go: super-easy, egg-free, cinnamon French toast.

- 3/4 cup (93 g) all-purpose flour
- 1 tablespoon (12 g) evaporated cane juice
- 1/4 teaspoon ground cinnamon
    Pinch salt
- 1 cup (235 ml) nondairy milk
- 6 to 8 slices of thick, stale bread

IN A SHALLOW BOWL or dish, combine flour, evaporated cane juice, cinnamon, and salt. Whisk in milk until smooth.

Preheat a nonstick pan or skillet over high heat.

Dip each slice of bread into the mixture to coat with a thin layer.

Place bread into the pan and "toast" for about 2 minutes per side, or until golden and crispy. Flip and repeat.

Repeat with remaining slices.

Serve hot, smothered in nondairy butter and dark maple syrup.

YIELD: 6 to 8 slices

Note: Stale bread is strongly recommended in this recipe. as fresh bread makes the toast soggy. If you don't have stale bread. fake it by putting slices of fresh bread on a baking sheet and placing it in the oven at the lowest setting for about an hour.

DROP IT LIKE IT'S A BISCU-IT

 Egg sub: vegetable purée (squash and lentil) for binding and moisture

Crispy outside, tender inside—the best of both worlds! The squash and lentil spread gives great flavor and body to this tasty treat. Enjoy these freshly baked with a big bowl of soup, and be merry.

2/3 cup (160 ml) unsweetened soymilk
1 cup (270 g) Roasted Squash and Lentil Spread (page 256)
1/3 cup (80 ml) peanut oil
1 teaspoon fine sea salt
1 teaspoon granulated onion
1 teaspoon ground coriander
1 teaspoon red pepper flakes to taste
2 1/2 cups (300 g) whole wheat pastry flour
1 tablespoon (12 g) baking powder

PREHEAT OVEN to 375°F (190°C, or gas mark 5). Line a large cookie sheet with parchment paper or a silicone baking mat.

Combine milk, spread, oil, salt, onion, coriander, and pepper flakes in a large bowl; mix thoroughly.

Sift flour and baking powder on top of mixture. Stir well until combined.

Scooping out a generous 1/3 cup (134 g) preparation per biscuit, place dough on prepared sheet.

Bake for 30 minutes, or until biscuits are golden brown.

Let cool on a wire rack. Enjoy warm from the oven, or toast before eating.

YIELD: 6 biscuits

CRANBERRY BANANA BREAKFAST CAKE

 Egg sub: fruit purée (banana) for binding

Bananas deliver binding power and moisture to this lightly sweetened cake that makes a great breakfast item when yet another bowl of oatmeal just won't do.

Nonstick cooking spray
2 cups (240 g) whole wheat pastry flour
2 teaspoons baking powder
1/4 teaspoon fine sea salt
3 small (1 cup, or 240 g) ripe bananas
1/2 cup (96 g) raw sugar
1 teaspoon pure vanilla extract
3/4 cup (180 ml) brewed coffee, cooled
3 tablespoons (45 ml) peanut oil
1 cup (95 g) fresh cranberries
1/3 cup (40 g) chopped walnuts
1/3 cup (67 g) vegan white chocolate, carob, or semisweet chocolate chips

PREHEAT OVEN to 350°F (180°C, or gas mark 4). Lightly coat a 9-inch (23 cm) round baking pan with cooking spray.

In a large bowl, combine flour, baking powder, and salt.

In a medium bowl, mash bananas into small chunks along with sugar, vanilla, coffee, and oil.

Toss cranberries, walnuts, and chips with the flour mixture.

Fold wet ingredients into dry, being careful not to overmix.

Pour batter into prepared pan and bake for 35 minutes, or until a toothpick inserted in the center comes out clean.

Let cool on wire rack for 15 minutes before removing from pan and cooling completely.

YIELD: 6 to 8 servings

CARROT CAKE WAFFLES

 Egg sub: vegetable purée (carrot) for binding

The puréed carrots help bind these hearty waffles, to which you can add shredded coconut, raisins, or fresh pineapple. Serve with Basic Maple Cream (page 26).

    1 pound (454 g) carrots, peeled and chopped into 1-inch (2.5 cm) pieces
    1 cup (80 g) old-fashioned oats
    1 cup (117 g) toasted walnut halves
    1 cup (120 g) light spelt flour
    1 tablespoon (7 g) ground cinnamon
    1 teaspoon ground ginger
 $1/2$ teaspoon grated nutmeg
 $1/2$ teaspoon fine sea salt
    2 teaspoons baking powder
 $3/4$ cup (144 g) Sucanat
 $1/2$ cup (120 ml) fresh orange juice
$11/2$ cups (355 ml) light coconut milk
 $1/4$ cup (60 ml) melted coconut oil
    2 teaspoons pure vanilla extract
       Nonstick cooking spray

PLACE CARROTS in a medium saucepan; add enough water to cover by 1 inch (2.5 cm). Cook for 30 minutes, or until carrots are tender enough to purée. Let cool completely before using.

Combine oats, walnuts, flour, spices, salt, baking powder, and Sucanat in a food processor. Process until finely ground. Transfer to a large bowl.

Place cooled carrots, orange juice, $1/2$ cup (120 ml) of milk, oil, and vanilla in food processor. Process until puréed.

Add purée to dry ingredients. Add remaining 1 cup (235 ml) milk as needed, until the batter is pourable without being too thin.

Follow waffle iron instructions to prepare waffles, remembering to spray iron between each waffle for best results.

YIELD: 8 Belgian or 16 standard waffles

BANANA CHUTNEY QUICK BREAD

 Egg sub: fruit purée (banana chutney) for binding

 Soy Free

This bread takes no time to throw together and will shine when served alongside soups or salads. The chutney is what replaces the egg in this recipe, giving it texture and tastiness.

> Nonstick cooking spray
> 1 cup (240 g) Banana Chutney (page 245)
> 2 teaspoons garam masala
> 1/4 teaspoon red pepper flakes
> 1/2 cup (128 g) crunchy natural peanut butter
> 2 tablespoons (30 ml) vegetable oil
> 2 cups (470 ml) water, up to 1/4 cup (60 ml) more if needed
> 1 1/2 cups (188 g) all-purpose flour
> 1 1/2 cups (180 g) white whole wheat flour
> 1 tablespoon (12 g) baking powder
> 1 1/2 teaspoons fine sea salt

PREHEAT OVEN to 350°F (180°C, or gas mark 4). Lightly coat an 8-inch (20 cm) square pan with cooking spray.

Combine chutney, garam masala, red pepper flakes, peanut butter, and oil in a medium bowl. Stir until emulsified.

Add water and stir until combined.

In a large bowl, whisk flours, baking powder, and salt.

Fold wet ingredients into dry, being careful not to overmix. Add extra water (1 tablespoon [15 ml] at a time) if the batter is too thick, as it will depend on the texture of both the chutney and peanut butter used.

Pour batter into prepared pan. Bake for 60 minutes, or until a toothpick inserted in the center comes out clean.

Let cool on a wire rack for at least 15 minutes before removing from pan.

The cake will look like it isn't done if you slice it before it's cooled.

YIELD: 6 to 8 servings

SPELT FLAX SAVORY QUICK BREAD

 Egg sub: soymilk, vinegar, and baking soda combination for leavening, fruit purée (apple) for moisture

 Corn Free

This versatile and rather dense quick bread is reminiscent of corn bread, even though it contains no corn. Intriguing, but we won't take too much time thinking about it and will gladly enjoy another slice instead.

> Nonstick cooking spray
> 1 tablespoon (15 ml) apple cider vinegar
> 1 cup (235 ml) soymilk
> 2¼ cups (315 g) whole spelt flour
> ¼ cup plus 2 tablespoons (42 g) flaxseed meal
> 1 teaspoon fine sea salt
> 2 teaspoons baking soda
> 2 tablespoons (30 ml) melted coconut oil
> 4 ounces (112 g) unsweetened applesauce

PREHEAT OVEN to 350°F (180°C, or gas mark 4). Lightly coat an 8 x 4-inch (20 x 10 cm) loaf pan with cooking spray.

Combine vinegar and soymilk in a medium bowl; it will curdle and become like buttermilk.

In a large bowl, combine flour, flaxseed, salt, and baking soda.

Add coconut oil and applesauce to the buttermilk mixture.

Pour batter into prepared pan.

Bake for 45 minutes, or until a toothpick inserted in the center comes out clean.

Place on a wire rack for 15 minutes before removing from pan; let cool completely before slicing or storing.

YIELD: One 8-inch (20 cm) loaf

PB AND J PIE

 Egg sub: nondairy yogurt for binding

Turning America's favorite snack into pie-form makes this dessert perfect for Independence Day—and every other day! Using nondairy yogurt helps make it even more delectable and tender.

   1 **cup (256 g) crunchy natural peanut butter**
 3/4 **cup (180 g) nondairy yogurt, store-bought or**
      Basic Homemade Sorta Yogurt (page 20)
   1 **cup (192 g) raw sugar**
   2 **teaspoons pure vanilla extract**
   1 **cup (120 g) light or (140 g)**
      whole spelt flour
   1 **teaspoon baking powder**
   1 **teaspoon ground cinnamon**
      One 9-inch (23 cm) prebaked vegan pie shell
 1/2 **cup (160 g) all-fruit strawberry jam**

PREHEAT OVEN to 350°F (180°C, or gas mark 4).

With an electric mixer, cream peanut butter, yogurt, sugar, and vanilla.

In a separate bowl, combine flour, baking powder, and cinnamon.

Add dry ingredients into wet, stirring until just combined. The batter will be stiff.

Divide the batter in two equal parts and spread half into the center of the pie shell. Using a frosting spatula or rubber spatula, start from the center, and bring the batter to the edges, just like when frosting a cupcake.

Spread the jam on top of the first layer of batter.

Spread the rest of the batter on top of jam. Don't worry if the jam escapes around the edges; it will create jellylike bits once baked.

Bake for 30 minutes, or until pie is set and turns golden brown.

YIELD: 8 slices

WHERE DO YOU GET YOUR PROTEIN: CAKE

 Egg sub: fruit purée (apple) for binding

A healthy, chocolaty cake that is a bit reminiscent of honey cake, even though it is made sans honey. The protein comes from both the tempeh and almond butter.

FOR CAKES:

- 2 **tablespoons (30 ml) peanut oil**
- 8 **ounces (227 g) tempeh, crumbled**
- 2 **cups (384 g) Sucanat**
   Nonstick cooking spray
- 2 **cups (470 ml) nondairy milk**
- 1 **tablespoon (15 ml) pure vanilla extract**
- 1/4 **cup (64 g) creamy natural almond butter**
- 1/4 **teaspoon fine sea salt**
- 4 **ounces (112 g) unsweetened applesauce**
- 1 1/4 **cups (156 g) all-purpose flour**
- 1 **cup (120 g) whole wheat pastry flour**
- 2/3 **cup (53 g) unsweetened cocoa powder**
- 2 1/4 **teaspoons baking powder**

FOR TOPPING:

- 1 **cup (240 g) raspberry jam**
- 1/2 **cup (40 g) unsweetened cocoa powder**

TO MAKE THE CAKES: Add oil to a medium skillet; add tempeh and cook over medium heat for about 6 minutes, or until golden brown. Stir Sucanat into tempeh until combined and dissolved.

Preheat oven to 350°F (180°C, or gas mark 4). Lightly grease two 9-inch (23 cm) round cake pans with cooking spray.

In a blender, combine tempeh, milk, vanilla, almond butter, salt, and applesauce. Blend until perfectly smooth.

In a large bowl, sift together flours, cocoa, and baking powder.

Fold wet ingredients into dry, being careful not to overmix.

Divide batter equally between prepared pans.

Bake for 35 minutes, or until a toothpick inserted in the center of cake comes out clean.

Let cool completely.

TO MAKE THE TOPPING: Combine jam and cocoa in a medium bowl until smooth. Divide equally among both cooled cakes and spread in an even layer.

Place one cake on top of the other. Enjoy at room temperature.

YIELD: 10 to 12 servings

SPICY AND WARM CUPCAKES

 Egg sub: soymilk, vinegar, and baking soda combination for leavening

Further proof of the powers that result from combining nondairy milk and vinegar! Sprinkle with powdered sugar right before serving and you're sure to find bliss.

Nonstick cooking spray
- 1/2 cup (112 g) nondairy butter, softened
- 1 1/4 cups plus 2 tablespoons (325 ml) soymilk
- 1 tablespoon (15 ml) apple cider vinegar
- 3/4 cup (144 g) Sucanat
- 1 1/2 teaspoons ground ginger
- 1 teaspoon ground cinnamon
- 1/2 teaspoon ground nutmeg
- 1/4 teaspoon fine sea salt
- 2 cups (240 g) whole wheat pastry flour
- 1 teaspoon baking soda

PREHEAT OVEN to 350°F (180°C, or gas mark 4). Lightly coat a standard muffin tin with cooking spray.

Using a blender, combine butter, soymilk, vinegar, Sucanat, spices, and salt and blend until smooth.

Whisk flour and baking soda in a large bowl.

Fold wet ingredients into dry, being careful not to overmix. Divide batter among muffin cups.

Bake for 20 minutes, or until a toothpick inserted in the center of a cupcake comes out clean.

YIELD: 12 cupcakes

WALNUT CHOCOLATE BROWNIES

 Egg sub: pudding for binding

Walnuts give a great intensity and richness to these treats, while the pudding imparts moistness and extra flavor. Just the way we like our brownies!

Nonstick cooking spray
- 1 cup (117 g) walnut halves
- 1/2 cup (96 g) Sucanat
- 1/2 cup (96 g) raw sugar
- 1/4 cup (60 ml) canola oil
- 2 teaspoons pure vanilla extract
- 1 cup (250 g) Chocolate Pudding (page 36)
- 1 cup (125 g) all-purpose flour
- 1/2 cup (40 g) unsweetened cocoa powder
- 1/4 teaspoon baking powder
- 1/4 teaspoon fine sea salt
- 1/2 cup (88 g) semisweet chocolate chips

PREHEAT OVEN to 350°F (180°C, or gas mark 4). Lightly coat an 8-inch (20 cm) square pan with cooking spray.

Combine walnuts and sugars in food processor. Process until finely ground.

In a medium bowl, combine ground walnuts, oil, vanilla, and pudding.

In a large bowl, sift together flour, cocoa, baking powder, and salt.

Fold wet ingredients into dry, being careful not to overmix. Fold in chips.

Bake for 32 minutes, or until brownies are firm and the sides start pulling away from the pan.

YIELD: 12 servings

CHICK-O-LATE BROWNIE CAKE

 Egg sub: vegetable purée (chickpea) for binding

Rich? Check. Decadent? Pretty much. The chickpeas are the secret weapon in a dessert that sneaks in protein as well as healthy fiber.

Nonstick cooking spray
$1^1/3$ cups (242 g) semisweet chocolate chips
3 tablespoons (42 g) solid coconut oil
$^1/3$ cup (64 g) raw sugar
$^1/3$ cup (64 g) Sucanat
Pinch fine sea salt
2 teaspoons pure vanilla extract
1 cup (99 g) pecans halves
1 cup (125 g) all-purpose flour
$^1/2$ teaspoon baking powder
1 can (15 ounces, or 425 g) chickpeas, drained and rinsed

PREHEAT OVEN to 350°F (180°C, or gas mark 4). Lightly coat an 8-inch (20 cm) square pan with cooking spray.

Combine chips and coconut oil in a small bowl. Microwave for 1 minute.

Stir and microwave again if needed until chips melt.

Stir sugars, salt, and vanilla into the mixture. Set aside to cool.

Process $^2/3$ cup (66 g) of pecans until ground; add flour and baking powder and pulse a few times until incorporated. Add chickpeas, process until puréed, scraping sides once.

Add chocolate mixture to food processor and pulse until just combined, scraping sides once.

Add remaining $^1/3$ cup (38 g) pecans, pulse a few times until just combined. The batter will be stiff.

Pour batter into prepared pan. Bake for 25 minutes, or until set and the sides pull away from the pan.

Let cool on a wire rack before removing from pan. Let cool completely before enjoying, lest the brownies be crumbly.

YIELD: 8 to 10 servings

Note: Make the cake gluten free by using 1 cup (160 g) sweet rice flour instead of all-purpose flour, and increase baking powder to 1 teaspoon.

MARZIPAN-FILLED PILLOWS

 Egg sub: starch and water combination for binding

Marzipan is usually made with egg whites, but potato starch works beautifully here. You will have leftover marzipan for future baking projects—or just for nibbling.

FOR MARZIPAN:

  1 cup (96 g) almond meal
1¹/₂ teaspoons potato starch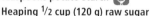
    Heaping ¹/₂ cup (120 g) raw sugar
 ¹/₂ teaspoon pure almond extract
1¹/₂ tablespoons (23 ml) water

FOR PILLOWS:

1¹/₃ cups (240 g) chopped bittersweet chocolate
  1 cup (120 g) whole wheat pastry flour
 ¹/₂ teaspoon baking powder
    Pinch fine sea salt
    Packed ¹/₄ cup (60 g) organic brown sugar
 ¹/₄ cup (60 ml) peanut oil
  2 teaspoons pure vanilla extract
  2 tablespoons (30 ml) water, more if needed
    Heaping ¹/₂ cup (120 g) homemade marzipan

TO MAKE THE MARZIPAN: Combine almond meal, starch, and sugar in food processor. Process until combined. Add almond extract and water, until paste forms.
  Wrap in plastic; store in fridge.

TO MAKE THE PILLOWS: Preheat oven to 350°F (180°C, or gas mark 4). Line 2 cookie sheets with parchment paper or silicone baking mats.
  Combine the chocolate, flour, baking powder, salt, and brown sugar in food processor. Process until finely ground.
  Add oil and vanilla; pulse until combined. Add water until dough holds together when pinched.
  Divide dough into 16 equal portions of 1 heaping tablespoon (32 g) each.
  Shape and flatten dough into disks. Divide marzipan into 8 equal portions of 1 heaping tablespoon (15 g) each.
  Place marzipan in the center of 8 cookie dough disks. Cover with another cookie dough disk, pressing down and flattening dough to seal the edges. If the dough starts to dry out from being handled, moisten your hands with water. Be careful not to make the edges too thin or they will bake too quickly and burn.
  Bake for 12 minutes, or until set. Leave on cookie sheet for 10 minutes. Let cool on wire rack.

YIELD: 8 large cookies and about ½ cup (120 g) leftover marzipan

WALNUT COOKIES

 Egg sub: coconut paste for binding

 Soy Free

As we were cleaning the coffee grinder after working on another project, we noticed the water turned the ground coconut into a rather interesting paste that could be used as an egg replacer.

1 cup (94 g) shredded coconut, ground into fine flour

1/4 cup plus 2 tablespoons plus 2 teaspoons (100 ml) hot water

1 cup (117 g) walnut halves, half turned into butter (see page 251 for more info on making nut butters), other half coarsely chopped

1 cup (192 g) Sucanat

1/4 cup (60 ml) peanut oil

1 teaspoon ground ginger

1 teaspoon ground cinnamon

1/2 teaspoon ground cardamom

1/2 teaspoon ground allspice

2 teaspoons pure vanilla extract

1/2 teaspoon fine sea salt

1 1/2 cups (180 g) whole wheat pastry flour

1/2 teaspoon baking powder

PREHEAT OVEN to 350°F (180°C, or gas mark 4). Line 2 large cookie sheets with parchment paper or silicone baking mats.

Thoroughly combine coconut with water in a large bowl; the coconut flour will dissolve and form a paste.

Stir in walnut butter and chopped walnuts, Sucanat, oil, spices, vanilla, and salt. Sift flour and baking powder on top, and stir until combined.

Divide dough using 2 heaping tablespoons (57 g) per cookie.

Flatten dough on prepared sheets; the cookies will not spread much while baking.

Bake for 15 minutes, or until cookies are golden brown around the edges.

Let cool on sheets for a few minutes until firm enough to transfer to a wire rack to cool completely.

YIELD: 12 large cookies

COOKIE COOKIES

 Egg sub: egg replacer powder for binding and leavening

We first witnessed the addition of Oreo pieces into a homemade cookie on cookiemadness. net and promptly ran into the kitchen to experiment with the idea. We're hooked!

$^1/_2$ cup (112 g) nondairy butter

1 cup (192 g) raw sugar, ground into fine powder

1 tablespoon (8 g) egg replacer powder, such as Ener-G

$^1/_4$ cup (60 ml) warm water

$^1/_2$ teaspoon fine sea salt

2 teaspoons pure vanilla extract

$2^1/_2$ cups (300 g) whole wheat pastry flour

$^1/_4$ cup plus 2 tablespoons (66 g) semisweet chocolate chips

$^1/_2$ teaspoon baking powder

Nondairy milk, if needed

10 store-bought vegan cream-filled cookies, quartered

PREHEAT OVEN to 350°F (180°C, or gas mark 4). Line 2 large cookie sheets with parchment paper or silicone baking mats.

Using an electric mixer, cream together butter and sugar until fluffy.

Whisk together egg replacer and water in a small bowl until foamy.

Add egg mixture to butter mixture, along with salt and vanilla.

Beat for another minute. Combine flour, chips, and baking powder in a separate dish. Add to butter mixture, and mix until combined. If the dough is too dry, add some milk until it is more manageable. Gently fold in cookie pieces.

Scooping approximately $^1/_4$ cup (75 g) of dough, shape 12 large cookies and place them on the prepared sheets. Flatten the dough as much as you want the cookies to be, as they won't spread much during baking.

Bake cookies for 16 to 18 minutes, until slightly golden around the edges. Remove from oven and let cool, until the cookies are firm enough to transfer to a wire rack.

YIELD: 12 large cookies

SWEET POTATO BARS

 Egg sub: vegetable purée (sweet potato) for moisture and binding

 Wheat Free

If you've always dreamed of eating a thick oatmeal chocolate cookie in bar form, your wish has been granted: The sweet potato and cornstarch combo helps bind this portable and irresistible treat.

Nonstick cooking spray
1¼ cups (112 g) shredded coconut
2 cups (160 g) old-fashioned oats
1 cup (192 g) raw sugar
½ cup (88 g) semisweet chocolate chips
¼ cup (32 g) cornstarch
1⅓ cups (325 g) sweet potato purée
¼ cup (56 g) coconut oil, melted
2 teaspoons pure vanilla extract
2 teaspoons ground cinnamon
½ teaspoon fine sea salt
2 teaspoons baking powder
Nondairy milk, if needed

PREHEAT OVEN to 350°F (180°C, or gas mark 4). Lightly coat an 8-inch (20 cm) square baking pan with cooking spray.

Combine coconut, oats, sugar, and chips in food processor.

Process until finely ground, for about 2 minutes.

In a separate large bowl, combine cornstarch, purée, oil, vanilla, and cinnamon.

Add oat mixture, along with salt and baking powder, into sweet potato mixture. Stir until combined; add milk if the batter is difficult to mix.

Place the batter into prepared pan. Bake for 30 minutes, or until golden brown. Cool in pan on a wire rack.

YIELD: 10 to 12 bars

CINNAMON OAT COOKIES

Egg sub: nondairy milk (full-fat coconut) for moisture

Coconut milk plays the egg role in these cinnamony treats by adding the richness eggs usually impart to traditional cookie recipes.

⅔ cup (149 g) nondairy butter
1½ cups (288 g) Sucanat
½ cup (120 ml) full-fat coconut milk
2 tablespoons (14 g) ground cinnamon
½ teaspoon fine sea salt
2 teaspoons pure vanilla extract
2½ cups (200 g) old fashioned oats
2½ cups (300 g) whole wheat pastry flour
2 teaspoons baking powder
⅔ cup (108 g) packed raisins or (117 g) semisweet chocolate chips

PREHEAT OVEN to 350°F (180°C, or gas mark 4). Line 2 large cookie sheets with parchment paper or silicone baking mats.

Using an electric mixer, cream together butter and Sucanat. Add coconut milk, cinnamon, salt, and vanilla. Beat for another minute.

In a separate bowl, combine oats, flour, and baking powder.

In 3 batches, add dry ingredients to the wet ingredients. Add raisins or chocolate chips. Stir until combined.

Scoop ⅓ cup (99 g) dough per cookie; place on prepared sheets.

Flatten slightly, as cookies won't spread much while baking.

Bake for 16 minutes, until cookies are golden brown at the edges.

Leave on sheets for at least 10 minutes before transferring to a wire rack to cool.

YIELD: 12 large cookies

SPECULOOS-SPICED COOKIES

 Egg sub: just naturally egg free!

Spicy and reminiscent of toffee, these cookies are so self-sufficient they need no help with binding. They are also used as a crust in Chocoplum Pie (page 236).

1/2 **cup (80 g) dry-roasted unsalted almonds**
1 **cup (192 g) Sucanat**
1/4 **cup (40 g) brown rice flour**
1 **cup (140 g) whole spelt flour**
1/2 **teaspoon fine sea salt**
1/8 **teaspoon ground white pepper**
1 **tablespoon plus 1 teaspoon (9 g) ground cinnamon**
1 **teaspoon ground nutmeg**
1 **teaspoon ground ginger**
1/2 **teaspoon baking powder**
1/2 **cup (112 g) nondairy butter**
2 **tablespoons (30 ml) pure maple syrup**

COMBINE ALMONDS, Sucanat, flours, salt, spices, and baking powder in food processor; process until finely ground.

Add butter and pulse several times, until well incorporated.

Add syrup and pulse a few times until combined. Add a few teaspoons of water, if the dough is too dry and doesn't hold together when pinched.

Chill dough for at least 1 hour. Preheat oven to 350°F (180°C, or gas mark 4). Line 2 large cookie sheets with parchment paper or silicone baking mats.

Scoop out 2 heaping tablespoons (40 g) of dough and place onto cookie sheet: the dough will spread while baking, so leave about 1 inch (2.5 cm) between each cookie.

Bake for 12 minutes, or until cookies turn golden brown around the edges.

Let cool on sheet until firm enough to transfer to wire rack to cool completely.

YIELD: 16 cookies

TAMARIND BANANA COOKIES

 Egg sub: nondairy milk (soy) and oil combination for moisture

 Corn Free

These have a soft, chewy consistency when eaten out of the fridge and a wonderfully salty edge. The soymilk and oil combine to replace the egg here.

1/2 **cup (112 g) nondairy butter**

1 **cup (192 g) Sucanat**

2 **tablespoons (30 g) tamarind paste**

1/2 **teaspoon lemon zest**

1/2 **teaspoon grated fresh ginger**

1/4 **cup (60 ml) soymilk**

1 **teaspoon neutral-flavored oil**

2/3 **cup (40 g) soft dried banana, cut into strips with scissors or diced**

1/2 **teaspoon fine sea salt**

2 **cups (240 g) whole wheat pastry flour**

1 **teaspoon baking soda**

PREHEAT OVEN to 350°F (180°C, or gas mark 4). Line 2 large cookie sheets with parchment paper or silicone baking mats.

Using an electric mixer, cream together butter and Sucanat. Add tamarind, zest, and ginger. Stir in milk, oil, banana, and salt. Add flour and baking soda.

Scoop up about 2 tablespoons (57 g) of dough and place on prepared sheets. No need to flatten the cookies, as they spread while baking.

Bake for 12 minutes, or until cookies set and the edges start turning golden brown.

Cool on cookie sheets until firm enough to transfer to wire rack. Enjoy cold from the fridge.

YIELD: 12 large cookies

GINGER APPLE COFFEE CAKE

 Egg sub: fruit purée (apple) for binding

Light, moist, and nutty: That's what this cake is all about. The recipe may look involved, but its wonderful autumnal flavor will reward you for your efforts.

FOR SIMPLE GINGER SYRUP:

- ³/4 **cup (144 g) raw sugar**
- ³/4 **cup (180 ml) spiced apple cider**
- ³/4 **cup (72 g) thick slices peeled fresh ginger**

FOR SIMPLE STREUSEL:

- ¹/2 **cup (50 g) pecan halves, coarsely chopped**
- ¹/4 **cup (48 g) Sucanat**
- 1 **teaspoon ground cinnamon**
- 2 **teaspoons vegetable oil**

FOR CAKE:

- **Nonstick cooking spray**
- **Scant ¹/2 cup (113 g) Homemade Applesauce (page 231)**
- ¹/4 **cup (60 ml) peanut oil**
- ³/4 **cup (105 g) whole spelt flour**
- 1 **cup (125 g) all-purpose flour**
- 2 **teaspoons baking powder**
- 2 **teaspoons ground cinnamon**
- ¹/2 **teaspoon fine sea salt**
- 1 **cup (135 g) roasted apples from unblended, no-juice-added Homemade Applesauce (page 231), chopped a bit smaller**
- ¹/3 **cup (45 g) finely chopped candied ginger**

TO MAKE THE GINGER SYRUP: Combine all ingredients in a saucepan, bring to a boil and simmer for 20 minutes or until syrup has reduced to 1 cup (235 ml). Let cool; discard ginger.

TO MAKE THE STREUSEL: Combine all ingredients in a small bowl. Set aside.

TO MAKE THE CAKE: Preheat oven to 350°F (180°C, or gas mark 4). Lightly coat an 8-inch (20 cm) square pan with spray.

In a medium bowl, combine applesauce, oil, and ginger syrup. In a large bowl, sift together flours, baking powder, cinnamon, and salt.

Fold wet ingredients into dry, being careful not to overmix. Fold in apples and candied ginger. Pour batter into prepared pan. Sprinkle streusel evenly on top. Bake for 35 minutes or until cake is golden brown and firm in the center.

Transfer to a wire rack and cool before removing from pan.

YIELD: 8 to 10 servings

SPICY RED BUFFALO OR SPICY GREEN VERDE MACARONI SALAD

 Egg sub: silken tofu for mayo

Maybe Joni's been working in a deli for too long, but she can always be counted on to whip up a great macaroni salad for get-togethers.

This salad works two ways—green or red—depending on what kind of mood you're in! Both have a spicy kick and are mayo free.

FOR RED BUFFALO DRESSING:

- **14 ounces (397 g) soft silken tofu, drained but not pressed**
- **¹/₂ cup (112 g) nondairy butter, melted**
- **¹/₄ cup (60 ml) vegetable oil**
- **2 to 4 tablespoons (30 to 60 ml) hot sauce to taste**
- **1 tablespoon (8 g) garlic powder**
- **Salt and pepper to taste**

FOR GREEN SPICY VERDE DRESSING:

- **14 ounces (397 g) soft silken tofu, drained but not pressed**
- **¹/₃ cup (80 ml) Aji Verde (page 208)**
- **1 tablespoon (8 g) garlic powder**
- **Salt and pepper to taste**

FOR SALAD:

- **1 pound (454 g) elbow macaroni, prepared according to package instructions**
- **1 cup (30 g) fresh baby spinach**
- **1 cup (170 g) chopped roasted red peppers**
- **¹/₂ cup (80 g) diced red onion**
- **1 cup (100 g) chopped celery**
- **1 tablespoon (8 g) red pepper flakes, to taste**

TO MAKE THE DRESSINGS: Put all ingredients in a blender and purée until smooth.

TO MAKE THE SALAD: Rinse and drain cooked pasta and return to the pot. Add dressing, spinach, roasted red peppers, onion, celery, and red pepper flakes. Toss to coat. Serve warm or cold.

YIELD: 8 main-dish or 12 to 14 side-dish servings

CAKELIKE CORN BREAD

 Egg sub: fruit purée (apple) for binding

The alcohol in this recipe is so subtle that you can barely tell it's there. It just gives the dish a little something-something that makes for a scrumptious treat.

This corn bread is cakelike not because of sweetness (so have at it with chilies and other savory stuff, too) but because of its delicate and refined texture. And no dryness to be found here—thanks to the applesauce!

Nonstick cooking spray
1¼ cups (175 g) cornmeal
½ cup (63 g) all-purpose flour
1 teaspoon fine sea salt
2 teaspoons baking powder
1 tablespoon (15 ml) tequila
1 tablespoon (15 ml) triple sec
1 tablespoon (21 g) agave nectar
¾ cup (180 ml) unsweetened nondairy milk
¼ cup (60 ml) peanut oil
4 ounces (112 g) unsweetened applesauce
2 tablespoons (30 g) fire-roasted diced green chilies, or ½ teaspoon red pepper flakes

PREHEAT OVEN to 375°F (190°C, or gas mark 5). Lightly grease an 8-inch (20 cm) square baking dish with cooking spray.

In a large bowl, whisk together cornmeal, flour, salt, and baking powder.

In a medium bowl, whisk together tequila, triple sec, agave, milk, oil, applesauce, and chilies. Fold wet ingredients into dry, being careful not to overmix.

Pour batter into prepared dish. Bake for 35 minutes, or until cornbread is golden brown, firm on top, and the sides have started to pull away from the pan.

Let cool on a wire rack before slicing.

YIELD: 6 to 8 servings

Note: For a denser corn bread, add an extra ½ cup (63 g) flour to the batter. The corn bread will be ready in 25 minutes.

KEEP THE FARM ANIMALS
FLOURISHING!

FOOLPROOF
SUBSTITUTIONS
FOR MEAT

Chapter 4

From Savory Burgers to Black Forest BLTs: HOW TO SUBSTITUTE FOR BEEF, CHICKEN, PORK, AND SEAFOOD

SO HOW WILL YOU GET ENOUGH PROTEIN when you take meat out of the picture?

Let's face it, protein is essential for a healthy life: We need it to build strong bodies, hormones, grow muscle, bones, nails, hair. . . . you get the idea. Protein is created by amino acids, some of which are naturally created by our bodies, while the rest needs to come from what we eat.

With that in mind, it is actually quite easy to create hearty, delicious, and protein-packed meals with all the tasty foods that can be harvested naturally from the earth, like beans, whole grains, quinoa, nuts, seeds, legumes, and soy.

Consider the Facts

You are what you eat, and unfortunately, so is your meat, pork, chicken, and fish. Because of factory farming and pollution, the meat sold at most markets today is not the same as that which our grandparents served up for dinner. Hormones, pesticides, chemicals, and pollution run rampant in today's factory farmed meats. If that is not enough to make you stop eating meat, consider this:

- **Beef:** Saturated fats in red meat cause heart disease, atherosclerosis, colon cancer, rheumatoid arthritis, and endometriosis. And that's what it does to your body. Shall we go into detail about what it does to the cow's body?

- **Poultry:** Potentially fatal, salmonella poisoning is far more prevalent in poultry than any other food. *Consumer Reports* found over two-thirds of store-bought chicken contained measurable levels of this harmful bacteria. Oh, and did we mention the battery cages?

- **Pork:** In addition to having many of the same health concerns as beef and poultry, pork is high in histamine levels as well as sulfur. Also, pigs in large factory farms are often fed the worst of foods, which eventually find a way to your plate. Besides, did you know that pigs are known to be even smarter than the average companion dog?

- **Seafood:** Pollution has produced many chemicals, including high levels of mercury, that flow through the fish in our oceans. There is a website that publishes "The Fish List," which lists the best and worst seafood to buy based on food safety and environmental issues. If you need a list to tell you if your food is safe, don't you think it's time to switch to a new food?

For more information on the horrors of factory farming, visit www.meat.org.

Guidelines for Substituting for Meat

Recreating traditionally meat-based dishes is actually a pretty easy task. As you begin to experiment with different proteins, you will find which ones you prefer and how to season them to mimic the tastes and textures of the meat you are trying to replace.

Generally speaking, when subbing a vegan protein for an animal-based equivalent in a recipe, use the same weighted measure. For instance, if a recipe calls for 6 ounces (170 g) of beef, use 6 ounces (170 g) of a beef-style or plain seitan cutlet.

For more specific guidelines, look at the following chart, and then we'll apply our knowledge to a nonvegan recipe.

IF THE ORIGINAL RECIPE CALLS FOR...	REPLACE WITH...
8 ounces (227 g) bacon (bits)	• 8 ounces (227 g) store-bought imitation bacon bits, such as Bacuns • 8 ounces (227 g) Imitation Bacon Bits (page 143)
8 ounces (227 g) bacon (strips)	• 8 ounces (227 g) store-bought vegan bacon, such as Lightlife Smart Bacon • 8 ounces (227 g) Black Forest Bacon (page 142)
1 cup (235 ml) beef broth	• 1 cup (235 ml) store-bought beef-flavored vegetable broth, such as Better than Bouillon No Beef Base • 1 cup (235 ml) plain vegetable broth • 1 tablespoon (15 ml) steak sauce and 1 tablespoon (15 ml) soy sauce mixed with 1 scant cup (205 ml) plain vegetable broth
8 ounces (227 g) beef (ground)	• 8 ounces (227 g) store-bought ground beef substitute, such as Lightlife Gimme Lean • 8 ounces (227 g) Textured Vegetable Protein (TVP) reconstituted with beef-flavored broth
8 ounces (227 g) beef (steaks or burgers)	• 8 ounces (227 g) store-bought veggie burgers, such as Wildwood or Morningstar • 8 ounces (227 g) Portobello mushrooms • 8 ounces (227 g) Baked Seitan Cutlets (page 130) • 8 ounces (227 g) Basic Traditional Boiled Seitan (page 110)
8 ounces (227 g) beef (strips)	• 8 ounces (227 g) Basic Traditional Boiled Seitan (page 110) cut into strips • 8 ounces (227 g) Portobello mushrooms cut into strips
8 ounces (227 g) chicken (breasts)	• 8 ounces (227 g) store-bought products, such as Gardein or Match Foods • 8 ounces (227 g) Baked Seitan Cutlets (page 130)
1 cup (235 ml) chicken broth	• 1 cup (235 ml) chicken-flavored vegetable broth, or plain vegetable broth
8 ounces (227 g) chicken (strips or nuggets)	• 8 ounces (227 g) chicken-style store-bought strips or nuggets, such as Gardein or Morningstar
8 ounces (227 g) chicken or turkey (ground)	• 8 ounces (227 g) TVP reconstituted in chicken-flavored broth

(continued on next page)

(continued from previous page)

IF THE ORIGINAL RECIPE CALLS FOR...	REPLACE WITH...
8 ounces (227 g) chorizo	• 8 ounces (227 g) store-bought soy chorizo, such as Soyrizo • 8 ounces (227 g) Seitan Chorizo Crumbles (recipe included in *500 Vegan Recipes*)
8 ounces (227 g) crab (lump)	• 8 ounces (227 g) store-bought imitation crab, such as Match Foods
8 ounces (227 g) sliced deli meats	• 8 ounces (227 g) store-bought vegan deli meat slices, such as Tofurky or Yves • 8 ounces (227 g) Basic Traditional Boiled Seitan (page 110) thinly sliced
8 ounces (227 g) fish (fillets)	• 8 ounces (227 g) Fish-y Sticks (page 145) • 8 ounces (227 g) White Bean Cutlets (page 220)
8 ounces (227 g) hot dogs	• 8 ounces (227 g) store-bought vegan hot dogs, such as Tofurky or Tofupups • 8 ounces (227 g) All-American Hot Dogs (page 136)
8 ounces (227 g) pepperoni	• 8 ounces (227 g) store-bought vegan pepperoni, such as Yves • 8 ounces (227 g) TVP or Seitan Pepperoni
8 ounces (227 g) canned tuna	• 8 ounces (227 g) store-bought mock tuna, such as Tun-o • 8 ounces (227 g) Happy Sea Tempeh Salad (page 146)

VEGANIZED!: SAMPLE RECIPE

Let's have a look at the following traditional recipe to see how we would replace the meat and other nonvegan ingredients:

STEAK WITH SWEET PEPPER SAUCE

This meaty main course was adapted and veganized from the *Better Homes and Gardens New Cook Book.*

1 medium green sweet pepper, cut into thin strips

1 medium red sweet pepper, cut into thin strips

1/2 cup (80 g) diced onion

2 cloves garlic, peeled and minced

1 tablespoon (15 ml) cooking oil

> We would use peanut oil because its high smoke point makes it suitable for longer cooking times

1 pound (454 g) beef ribeye or top sirloin steak

Salt and pepper to taste

> The easiest substitute here would be 4 portobello mushrooms with stems removed. However, 4 Baked Seitan Cutlets (page 130) would also be an excellent choice

1/2 cup (120 ml) beef broth

1 tablespoon (3 g) fresh oregano or basil

> We would use vegetable broth, or 1 1/2 teaspoons Better Than Bouillon No-Beef Base mixed with 1/2 cup (120 ml) water

1 cup (250 g) chopped, seeded tomatoes

ADD OIL to a large skillet over high heat and sauté peppers, onion, and garlic for about 4 minutes, or until vegetables are still crisp but tender.

Remove vegetables from pan and set aside.

Reduce heat to medium; add portobellos or seitan cutlets to the same pan, adding more oil if necessary.

Add salt and pepper. Cook for about 4 minutes per side and transfer to a serving platter.

Add broth and oregano or basil to the pan. Bring to a boil; lower heat. Simmer uncovered for about 3 minutes, or until liquid is reduced by half.

Add cooked vegetables and tomatoes. Cook until just heated through.

Spoon mixture over portobellos or seitan cutlets.

YIELD: 4 servings

Finding Meat Substitutes at the Store

We can't imagine how vegans of years past could function without all of the amazing products now available.

If you have access to a local international food market, you will find some astonishingly realistic mock meats from chicken, to duck, to ham, even foods like mock crab and tuna. Two things to keep in mind, however: These foods are usually gluten-based, and as with all foods, moderation is key since many of these ready-made products are highly processed.

Breakfast patties and **sausages:** Lightlife steals the show for quick meaty breakfasts. Their Gimme Lean Sausage and Smart Links make an easy task of morning meals.

Deli slices: Great for lunches and picnics, Tofurky slices earn our praise for their taste and texture. These don't really taste like meat, but they do the trick when you need something quick to throw between two slices of bread.

Fake bacon: We love to make our own, in order to choose the strength of flavors, but we also rely on Lightlife with its Smart Bacon and Bacon Style Strips. When it comes to bacon-flavored bits, there are many to choose from; just make sure to check ingredient labels and avoid those with pesky hydrogenated fats.

Gardein: Oh, Gardein. You look and taste just like chicken! There are prepackaged dishes prepared with Gardein, and you may also find entire "breasts" in the deli department of your local health-food store.

Hot dogs, brats, and **veggie burgers:** These can be lifesavers when invited to a party, barbecue, or other social event, and almost every grocery store carries at least one variety of each. Double-check ingredients, as certain brands sneak in egg or dairy. Some of our favorites? Tofurky makes excellent wieners, and Wildwood makes some of the best veggie burgers in the world!

Pairing beans and **whole grains, nuts,** and **seeds:** Why not take a break from processed meat substitutes? Combine beans with whole grains, nuts, or seeds, and you get a complete protein! If you don't have the opportunity to pair these foods together in one single meal, you need not fret: It is actually still possible to obtain a complete protein by enjoying grains, beans, nuts, and seeds over the course of one to two days. They are especially tasty when served together though, so it really is up to you.

A few examples: whole-grain pita chips dipped into hummus, brown rice and beans, peanut butter on whole wheat bread, corn chowder enjoyed with sunflower-seed buttered toast. The possibilities are limitless and far less processed than most of the meat subs one finds at the store. See Beet Rice Salad (page 216) and Triple-B Stew (page 211) for ideas.

Seitan: There are several ready-made varieties available on the market. You can usually find them in the refrigerated section of your local health-food store, near the tofu.

Soy chorizo: We are particularly fond of both the Soyrizo brand and the soy chorizo sold at Trader Joe's. It's perfectly seasoned, and the texture is spot-on.

Soy Curls: Similar to TVP, but way, way bigger. These can be hard to find on grocery-store shelves, so if you'd like to give them a try, visit Butlerfoods.com to order online, or to find a store near you.

Tempeh: There are many varieties of this fermented treat from plain soy to multigrain. We suggest you start with a plain variety and then experiment with different flavors until you find one you love. Tempeh comes either vacuum-sealed (almost always pasteurized) and more rarely, fresh (unpasteurized). Simmering vacuum-sealed tempeh in 2 cups (470 ml) vegetable broth, along with a dried bay leaf, for 20 minutes helps cut out the bitterness. Drain well and use in any recipe. Unpasteurized tempeh must be boiled for 20 minutes before eating.

Tofu: The silken variety comes in soft, medium, firm, and extra firm. The regular version comes in soft, medium, firm, extra firm, and even super firm. The latter is fantastic when fried, because it doesn't need to be pressed and yields deliciously crispy bits. Silken is best when making sauces and desserts, although we've had great results using regular firm tofu in mousses and pies.

TVP (Textured Vegetable Protein): A versatile meat substitute that has the appearance of Grape Nuts in its smaller granulated form. When reconstituted with vegetable broth or water, it mimics the texture of ground meats. It has very little flavor on its own and, much like tofu or seitan, takes on the flavor of whatever it is prepared with. Very low in fat, high in protein and fiber, it is produced from defatted soy flour and cooked under pressure, before being dried. We like Bob's Red Mill brand but also buy it from the bulk bins to save a buck. Note that TVP chunks come in various sizes, and the larger ones are handy when making curries, stir-fry dishes, and stews.

Vital wheat gluten: This flour will become a staple in your pantry from this point forward. It has endless uses in the vegan kitchen, from creating flavored meat substitutes (or seitan) from scratch to improving the texture and flavor of yeast bread.

Making Meat Substitutes at Home

Yes, there are a lot of ready-made meat substitutes right at your fingertips. But they can't match the satisfaction you gain from making your own from scratch. Tofu and tempeh are excellent sources of protein and are used extensively in the vegan kitchen. Freshly made artisan tofu and tempeh have a richer, nuttier, and often beanier flavor than their supermarket counterparts. And while it is possible to make fresh tofu and tempeh in the home kitchen, both require sterile workspaces and specialized equipment, so for our own safety, we prefer to purchase our tofu and tempeh.

The Recipes: Meat-Based Recipes...without the Meat!

Here we share the secrets to "beefing" up your repertoire of basic, made-from-scratch meat substitutes. And then we put the basics to good use to create hearty meat-free, protein-packed, savory dishes.

BASIC TRADITIONAL BOILED SEITAN

Once you master the basics of homemade seitan, you can experiment with different spice mixtures and techniques to make any number of variations.

FOR BROTH:

- 2.5 quarts (2.35 l) water
- 2 cups (470 ml) soy sauce
- 10 cloves garlic, peeled and halved
- 5 bay leaves
- Three 2-inch (5 cm) slices fresh ginger

FOR DOUGH:

- 1 cup (144 g) vital wheat gluten flour
- 5 cups (600 g) whole wheat flour
- 2$^1/_2$ cups (588 ml) water
- 3 stalks green onion, whites only, finely chopped
- $^1/_2$ cup (30 g) fresh chopped parsley
- 1 teaspoon garlic powder
- 1 teaspoon onion powder
- 1 teaspoon to 1 tablespoon (2 to 6 g) ground black pepper to taste
- Cheesecloth

COMBINE BROTH ingredients in a pot and bring to simmer while preparing dough.

In large bowl, mix flours together; slowly add water to form stiff dough. Knead about 70 times, right in the bowl if you'd like, and let rest 20 minutes.

Bring dough in bowl to the sink and cover with water. Knead the dough until the water becomes milky; discard water. Cover with fresh water and repeat process 10 to 12 times.

After last rinse, the texture of the dough will be wet and loose.

Add onions, parsley, and spices. Mix thoroughly by hand.

Divide dough in half. Place one half of dough in the center of a large piece of cheesecloth and roll tightly into a log. Tie ends to secure. Repeat with second half.

Place both logs in broth; simmer for 90 minutes.

Remove logs from broth and set on a plate to cool. Discard cheesecloth. (If cloth sticks, run under water and it will come off easily.)

Store seitan in fridge in airtight container. To retain moisture, add some broth to container. Keeps well for up to 2 weeks, or indefinitely in the freezer.

YIELD: About 4 pounds (1.8 kg)

SALISBURY STEAK WITH MUSHROOM, ONION, AND SAGE SAUCE

 Meat sub: TVP and vital wheat gluten

Like a TV dinner—that's made from scratch!

FOR STEAKS:

- 1 cup (235 ml) water or vegetable broth
- 1 cup (96 g) TVP granules
- 2 to 3 tablespoons (30 to 45 ml) soy sauce or tamari
- 1 cup (144 g) vital wheat gluten flour
- 1 teaspoon garlic powder
- 1 teaspoon onion powder
- 1/4 teaspoon paprika
- 1/3 cup (80 ml) olive oil
- Salt and pepper to taste

FOR SAUCE:

- 2 tablespoons (30 ml) olive oil
- 1 1/4 ounces (113 g) button mushrooms, sliced
- 3/4 cup (113 g) chopped white onion
- 2 tablespoons (30 g) minced garlic
- 2 tablespoons (30 ml) soy sauce or tamari
- 1 cup (235 ml) vegetable broth
- 1/4 cup (30 g) all-purpose flour mixed with 1/2 cup (120 ml) water to make a slurry
- 1/2 teaspoon powdered sage
- 1/4 cup (15 g) fresh parsley, finely chopped
- Salt and pepper to taste

TO MAKE THE STEAKS: In a microwave-safe bowl, combine water or broth, TVP, and soy sauce or tamari. Cover tightly with plastic wrap and microwave for 5 to 6 minutes. (Alternatively, bring water to a boil, pour over TVP and soy sauce, cover and let stand for 10 minutes to reconstitute.) Allow to cool.

Combine remaining ingredients with reconstituted TVP, using hands and mashing it together so the TVP doesn't crumble when shaping steaks. Let stand for 20 minutes to let gluten develop.

Divide into 6 equal portions, shaping into 1/4-inch (6 mm) -thick patties.

Preheat large skillet over medium heat. Fry steaks for 5 to 7 minutes per side, until golden brown and crispy. (Alternatively, bake them at 350°F [180°C, or gas mark 4] for 15 minutes.)

Serve topped with sauce.

TO MAKE THE SAUCE: Preheat oil in pan over medium-high heat.

Sauté mushrooms and onion for 5 to 7 minutes. Add garlic and cook for 2 minutes.

Add soy sauce or tamari and broth. Bring to a simmer. Stir in slurry and stir to thicken. Remove from heat.

Stir in sage, parsley, salt, and pepper.

Serve as is, or use blender to purée into a smooth sauce.

YIELD: 6 servings

Note: To get thin steaks, place one portion, slightly flattened, onto a plate. Place a second plate on top and apply pressure to flatten to desired thickness.

DIJON PORTOBELLO STEAKS

 Meat sub: portobello for steaks

 Corn Free Low Fat Nut Free

 Quick and Easy Soy Free  Wheat Free

There is nothing quite as naturally meaty without being meat as a thick juicy portobello steak! Serve the mushrooms with Yucca Fries (page 178) and punchy Aji Verde sauce (page 208), or use them to make an amazing panini with a schmear of Roasted Tomato Aioli (page 35). If you have leftover marinade, it works great as a salad dressing!

1/2 cup (120 ml) mild Dijon mustard
  3 tablespoons (45 ml) balsamic vinegar
  2 tablespoons (30 ml) agave nectar
     Salt and pepper to taste
  2 large portobello mushrooms, stems removed

IN A SMALL MIXING BOWL, whisk together mustard, vinegar, agave, salt and pepper.

   Place mushrooms in a shallow dish and cover with marinade. Let soak for at least 15 minutes.

   Remove mushrooms from marinade and place on the grill or in a grill pan; cook for about 7 minutes per side, basting with marinade, until mushrooms are tender.

YIELD: 2 servings

》 Clockwise from top: Aji Verde (page 208),
Dijon Portobello Steaks, Yucca Fries (page 178)

MEATFREE BALLS

 Meat sub: vital wheat gluten

 Corn Free

These are so delectable that you'll want to double or triple the batch so you can freeze some for next time. Serve in a sub-style sandwich, topped with your favorite slaw and dressing (as pictured), or top with your favorite marinara and shredded nondairy mozzarella—broil and *mangia*!

- 4 ounces (112 g) unsweetened applesauce, pumpkin purée, or tomato paste
- 2 tablespoons (30 ml) peanut oil
- 2 tablespoons (15 g) nutritional yeast
- 1/4 cup (36 g) vital wheat gluten
- 1 teaspoon onion powder
- 1/4 teaspoon garlic powder
- 1/2 teaspoon fine sea salt to taste
- 1/4 teaspoon ground black pepper to taste
- 1/2 cup (40 g) panko bread crumbs
- 1/4 cup (60 ml) BBQ sauce

PREHEAT OVEN to 375°F (190°C, or gas mark 5). Line a small baking sheet with parchment paper.

Combine the first seven ingredients in a medium bowl. Add bread crumbs and mix until well combined.

Divide and shape mixture into 8 balls. Place the balls in the center of the parchment paper on baking sheet; fold paper over balls to enclose them entirely. Arrange wrapped balls folded side down.

Bake for 20 minutes, or until golden brown. Unwrap and coat with BBQ sauce. Bake unwrapped for another 8 minutes.

YIELD: 8 1-inch (2.5 cm) meatfree balls

CARROT GINGER BURGERS

 Meat sub: vital wheat gluten and beans

 Corn Free

These meaty burgers are delicious served with Pear and Cauliflower Curry (page 32)!

- 3 large carrots, cut into bite-size chunks
- 2 cups (470 ml) fresh-squeezed orange juice
- 1/4 cup (60 ml) sesame oil (not toasted)
- 1/4 cup (60 ml) tamari
- 2 teaspoons garlic powder
- 1/4 cup (34 g) chopped candied ginger
- 1 1/2 tablespoons (23 ml) sriracha
- 3 tablespoons (45 ml) seasoned rice vinegar
- 1 can (15 ounces, or 425 g) chickpeas, drained and rinsed
- 3 cups (432 g) vital wheat gluten

COOK ALL INGREDIENTS except gluten in a saucepan over medium-high heat until the carrots are very tender, about 15 minutes. Purée mixture in a blender until smooth.

Combine with gluten: if you have a stand mixer, it helps to use it here, or use your hands to knead it in.

Preheat oven to 350°F (180°C, or gas mark 4). Line 2 large cookie sheets with parchment paper or silicone baking mats. Scoop 1/3 cup (95 g) of mixture, shaping and flattening a total of 17 burgers onto the prepared sheets.

Cover with foil to keep the burgers moist. Bake for 20 minutes on each side, or until golden brown. These burgers freeze well.

YIELD: 17 burgers

WESTERN BACON CHEESEBURGERS

 Meat sub: TVP and vital wheat gluten for ground beef, imitation bacon bits for bacon

Reminiscent of fast food burgers—only much healthier—these should be served with BBQ sauce, onion rings, and our BBQ Beans (page 156).

- 1 **cup (96 g) TVP granules**
- 1 **cup (235 ml) vegetable broth**
- 1/4 **cup (25 g) imitation bacon bits, store-bought or homemade Imitation Bacon Bits (page 143)**
- 1 **cup (144 g) vital wheat gluten**
- 1/8 **teaspoon liquid smoke**
- 1/2 **cup (60 g) nutritional yeast**
- 1 **tablespoon (8 g) garlic powder**
- 1 **tablespoon (8 g) onion powder**
- 2 **tablespoons (32 g) creamy no-stir nut butter**
- 2 1/2 **tablespoons (18 g) ground flaxseeds mixed with 3 tablespoons (45 ml) warm water**
- 1/4 **cup (60 ml) maple syrup**
- 1/4 **cup (60 ml) vegetable oil**
- 1/4 **cup (60 ml) BBQ sauce**
- **Salt and pepper to taste**

COMBINE TVP with broth in microwave-safe bowl, cover tightly with plastic wrap and microwave on high for 6 minutes. (Alternatively, bring broth to a boil, pour over TVP, cover and let stand for 10 minutes.) Let cool before handling.

Combine TVP and remaining ingredients in a medium mixing bowl and knead together for at least 5 minutes.

Let sit for a few minutes and form mixture into 4 patties.

Choose method of cooking: For grilling, cook for 5 to 7 minutes per side over a medium-high flame. For stovetop, sauté in oil for 5 to 7 minutes per side, or until browned and crispy. To bake, preheat oven to 350°F (180°C, or gas mark 4); place burgers on lined baking sheet and cook for 20 minutes; flip and bake for 20 more minutes. You can bake and then quickly pan-fry the burgers to crisp them up.

YIELD: 4 large (1/2 pound, or 227 g) burgers

TACO MEAT

 Meat sub: TVP for ground beef

 Quick and Easy

Use this Mexican spiced ground beef wannabe in tacos and burritos, on nachos, or pretty much anywhere you would want some yummy, spicy "beef."

FOR GLUTEN FREE VERSION:

- 1 **cup (96 g) TVP granules**
- 1 **teaspoon evaporated cane juice**
- 2 **teaspoons chili powder**
- 1 **teaspoon garlic powder**
- 1 **teaspoon onion powder**
- 1 **teaspoon ground cumin**
- 1 **teaspoon paprika**
- 1/2 **teaspoon cayenne pepper or ground chipotle powder**
- 1/2 **teaspoon salt**
- 1 **cup (235 ml) water**
- 2 **tablespoons (30 ml) olive oil (or other mild-flavored vegetable oil**

FOR SOY-FREE VERSION:

- 1 **cup (144 g) vital wheat gluten**
- 2 **teaspoons chili powder**
- 1 **teaspoon garlic powder**
- 1 **teaspoon onion powder**
- 1 **teaspoon Sucanat**
- 1 **teaspoon ground cumin**
- 1 **teaspoon paprika**
- 1/2 **teaspoon fine sea salt**
- 1/2 **teaspoon cayenne pepper or ground chipotle powder**
- 1/2 **cup (120 ml) water**
- 2 **tablespoons (30 ml) peanut oil**

IN A MICROWAVE-SAFE DISH, mix together TVP granules with evaporated cane juice, spices, and salt. Stir in water and oil. Cover tightly with plastic wrap and microwave on high for 5 to 6 minutes.

Alternatively, you can bring water to a boil, pour into TVP and spice mixture, add oil and stir to combine, cover, and let stand for 10 minutes.

Combine dry ingredients in a skillet, using a spatula. Add 1/4 cup (60 ml) water and oil. Using your fingertips, mix and crumble the mixture until no dry spots are left. Add a touch more water if mixture is too dry.

Cook crumbles over medium heat, breaking apart larger crumbles, and stirring often for 12 minutes, or until browned and cooked through.

YIELD: 2 1/2 cups (312 g) meat

CHILI CON CARNE

 Meat sub: vegetable based broth for beef base, TVP for ground beef

 Low Fat

If you're looking for something to slather on fries, hot dogs, baked potatoes, or corn chips, this bean-less and tomato-less chili is for you. Top it with diced onions and shredded vegan cheese, and oh, boy!

2$^1/_2$ cups (588 ml) water
   1 tablespoon (18 g) beef-flavored vegetable bouillon (such as Better than Bouillon No Beef Base)
$^1/_2$ cup (63 g) all-purpose flour mixed with 1 cup (235 ml) water to make a slurry
   1 cup (96 g) TVP granules
   2 tablespoons (16 g) chili powder
   2 tablespoons (14 g) dried minced onion
   1 tablespoon (7 g) paprika
   1 teaspoon garlic powder
$^1/_2$ teaspoon cayenne pepper, to taste
   1 teaspoon evaporated cane juice
   2 tablespoons (30 ml) white vinegar
      Salt and pepper to taste
$^1/_4$ teaspoon liquid smoke, optional

BRING WATER and bouillon to a boil in a pot; reduce to a simmer and slowly add flour slurry.

Whisk until thickened to the consistency of gravy.

Add TVP granules and continue to stir.

Stir in spices, evaporated cane juice, vinegar, salt, pepper, and liquid smoke, if using.

Cover and simmer for about 10 minutes.

YIELD: 4 cups (940 ml)

THIN PIZZA CRUST

 Meat sub: just naturally meat free!

This recipe yields a scrumptious pizza crust that's a perfect base for Sesame Orange Tempeh with Caramelized Onions Pizza (page 119) and Pineapple Tempeh Crumbles (with Pizza Option) (page 120).

$^1/_2$ cup (120 ml) Sesame Orange Marinade (page 119)
$^1/_2$ cup (120 ml) water
1$^1/_2$ teaspoons active dry yeast
   2 tablespoons (16 g) black sesame seeds
   1 tablespoon (8 g) brown sesame seeds
$^1/_2$ teaspoon fine sea salt
   1 cup (120 g) bread flour
1$^1/_2$ cups (180 g) white whole wheat flour
$^1/_2$ teaspoon peanut oil, to coat bowl

IN A MEDIUM BOWL, combine marinade and water.

Heat until lukewarm, about 100°F (38°C). Stir in yeast, and let foam for a few minutes.

In a large bowl, combine sesame seeds, salt, bread flour, and 1 cup (120 g) of white whole wheat flour. Stir wet ingredients into dry.

Transfer dough to a lightly floured surface and knead for about 8 to 10 minutes, until it is smooth and pliable, adding more wheat flour if dough is too wet.

Shape into a ball.

Lightly coat a large bowl with oil, place dough in bowl and turn to coat.

Cover tightly with plastic wrap, and let rise until doubled in size, 60 to 90 minutes.

To prepare pizza, follow directions in Pineapple Tempeh Crumbles (page 120) and Sesame Orange Tempeh with Caramelized Onions Pizza (page 119).

YIELD: 1 pizza crust

SESAME ORANGE TEMPEH WITH CARAMELIZED ONIONS PIZZA

 Meat sub: tempeh

For a satisfyingly complete meal, we like to eat this pizza with coleslaw.

One Thin Pizza Crust (page 118)

FOR SESAME ORANGE MARINADE:

3/4 cup (180 ml) orange juice

1 tablespoon (21 g) agave nectar

3 tablespoons (45 ml) sesame oil (not toasted)

3 tablespoons (45 ml) tamari

1 teaspoon sriracha to taste

1 clove garlic, peeled and minced

1/2-inch (1.25 cm) fresh ginger, peeled and minced

Pinch sea salt

FOR TOPPING:

8 ounces (227 g) tempeh, crumbled in small pieces

1/2 cup (120 ml) Sesame Orange Marinade

2 tablespoons (30 ml) peanut oil

1 large red onion, thinly sliced into half-moons

2 large shallots, thinly sliced

4 large cloves garlic, peeled and minced

1-inch (2.5 cm) piece fresh ginger, peeled and minced

1/2 teaspoon fine sea salt

3 tablespoons (45 ml) fresh lemon juice

1 teaspoon vegetable oil

2 tablespoons (18 g) cornmeal, to sprinkle sheet

PREPARE PIZZA DOUGH following recipe instructions.

TO MAKE THE MARINADE: Whisk together all ingredients.

TO MAKE THE TOPPING: Combine tempeh and marinade in medium bowl. Set aside. Heat peanut oil in saucepan. Add onion, shallots, garlic, and ginger. Cook over medium-high heat for 4 minutes, stirring often, until onions are tender. Add salt and lemon juice; cook until onions start to caramelize, about 10 minutes. Add tempeh and marinade; cook until tempeh browns nicely, about 6 minutes. Set aside.

TO ASSEMBLE PIZZA: Preheat oven to 400°F (200°C, or gas mark 6). Lightly brush a rimless 14 x 16-inch (36 x 41 cm) cookie sheet with oil and sprinkle with cornmeal. Punch down dough. Grab and stretch dough as thin as possible into a circle, being careful not to tear. Spread topping on dough. Let pizza rest for 10 minutes before baking. Bake for 24 minutes, until edges of crust turn golden brown. Cool on wire rack for several minutes before serving.

YIELD: 6 servings

PINEAPPLE TEMPEH CRUMBLES (WITH PIZZA OPTION)

 Meat sub: tempeh

Chickens join us in thanking you for choosing tempeh and want you to enjoy this dish on its own, alongside a healthy helping of brown jasmine rice or quinoa, or crumbled on top of a pizza crust.

> One Thin Pizza Crust (page 118), if making pizza
> 2 tablespoons (18 g) cornmeal, if making pizza
> 1 teaspoon vegetable oil, if making pizza
> 8 ounces (227 g) tempeh, crumbled
> 2 tablespoons (30 ml) pomegranate (or regular) red wine vinegar
> 1 tablespoon (15 ml) soy sauce
> 2 tablespoons (30 ml) coconut oil, melted
> 3 cloves garlic, peeled and minced
> 1 teaspoon garam masala
> 1 tablespoon (6 g) curry powder
> 1/2 teaspoon ground cumin
> 1 teaspoon mustard seeds, toasted
> 1 cup (156 g) fresh pineapple bits
> 1/3 cup (53 g) chopped onion
> 2 tablespoons (42 g) agave nectar
> 1/2 teaspoon red pepper flakes
> Generous handful fresh baby spinach, optional

IF MAKING PIZZA, prepare dough following recipe instructions.

While dough is rising, prepare the tempeh: Combine all ingredients in a large bowl, using your hands. Transfer to a large skillet and cook over medium-high heat, until onion is tender and crumbles start to brown, about 8 minutes.

Add spinach, if using, and cook until just wilted.

IF MAKING PIZZA: Preheat oven to 400°F (200°C, or gas mark 6). Lightly brush a rimless 14 x 16-inch (36 x 41 cm) cookie sheet with vegetable oil. Lightly sprinkle with cornmeal.

Punch down dough. Grab and stretch dough as thin as possible into a circle, being careful not to tear.

Add tempeh topping. Let stand for 10 minutes.

Bake for 22 minutes, until edges of crust turn golden brown.

Cool on wire rack for several minutes before serving.

YIELD: 6 servings

CHEESY "CHICKEN" CASSEROLE

 Meat sub: TVP or soy curls for pulled chicken, vegetable-based broth for chicken stock

This recipe calls for just about every category of "fake": fake cheese, fake meat, fake butter, fake sour cream. But don't be fooled. . . . It tastes *real* good!

- 1 **pound (454 g) penne pasta**
- 2 **cups (120 g) large chunk TVP or (80 g) Soy Curls**
- 2 **cups (470 ml) boiling water mixed with 2 tablespoons (16 g) "chicken-flavored" broth powder**
- 1/2 **cup (56 g) shredded nondairy cheddar**
- 1/2 **cup (56 g) shredded nondairy mozzarella**
- 1/4 **cup (56 g) nondairy butter**
- 1/3 **cup (80 g) nondairy sour cream**
- 1/4 **cup (60 g) nondairy cream cheese**
- 1 **cup (130 g) frozen peas**
- 1 **cup (130 g) frozen carrots**
- 1/3 **cup (27 g) panko bread crumbs**
  Salt and pepper to taste

PREHEAT OVEN to 350°F (180°C, or gas mark 4).

Boil the pasta in salted water, according to package directions. While pasta is cooking, reconstitute TVP by pouring boiling broth mixture over TVP, covering and allowing to sit for 10 minutes.

Drain pasta and return to pot.

Stir in shredded cheese, butter, sour cream, and cream cheese, until creamy. Stir in peas, carrots, reconstituted TVP (or Soy Curls), along with any remaining broth.

Transfer mixture to oven-safe 9 x13-inch (23 x 33 cm) casserole dish; sprinkle panko evenly on top.

Bake uncovered for 20 minutes. Increase heat to 450°F (230°C, or gas mark 8) and bake for another 10 minutes, or until browned on top.

YIELD: 8 servings

"CHICKEN" CORDON BLEU

 Meat sub: gardein or seitan for chicken breasts, vegan ham or
Black Forest Bacon for ham

This recipe can be made two ways. Super-duper easy using store-bought ingredients in a semi-homemade fashion, or from scratch using only homemade ingredients. Either way, we guarantee you will have a delicious meal worthy of guests or holiday dinners.

FOR BREADING:
- 1/2 **cup (63 g) all-purpose flour**
- 1 **cup (235 ml) unsweetened soymilk**
- 2 **tablespoons (30 ml) fresh lemon juice**
- 1 1/2 **cups (100 g) panko bread crumbs**
- 1 **tablespoon (2 g) dried parsley**
- 1 **teaspoon garlic powder**
- 1/4 **teaspoon paprika**
- **Salt and pepper to taste**

FOR SEMI-HOMEMADE VERSION:
- 4 **Gardein Chicken Breasts**
- 2 **ounces (56 g) vegan shredded cheese, such as Daiya**
- 4 **vegan ham-flavored deli slices, such as Yves**

FOR SCRATCH-MADE VERSION:
- 4 **Baked Seitan Cutlets (page 130)**
- 2 **ounces (56 g) Nutty Pepperjack (page 48), omitting peppers**
- 4 **slices Black Forest Bacon (page 142), without frying**

PREHEAT OVEN to 350°F (180°C, or gas mark 4). Line a baking sheet with parchment paper or silicone baking mat.

Place flour in a shallow dish.

Combine soymilk and lemon juice in a shallow bowl; it will curdle and become like buttermilk. In a shallow dish, mix together bread crumbs, parsley, garlic powder, and paprika.

Take each breast (or cutlet) and cut a slit in the side large enough to stuff with cheese and ham (or bacon).

Stuff 1/2 ounce (14 g) cheese and a slice of the ham (or bacon) inside the slit and seal with toothpicks.

Dip the stuffed breast or cutlet in flour to lightly coat; then dip into soymilk mixture; then into bread crumb mixture to coat.

Place breast or cutlet on baking sheet.

Repeat with remaining 3 breasts or cutlets.

Bake for 45 minutes, or until golden and crispy. Top with extra cheese during last 10 minutes of baking.

YIELD: 4 serving

TOFFALO HOT WINGS WITH COOL RANCH DIPPING SAUCE

 Meat sub: tofu, seitan, or tempeh

Enjoy this spicy finger food while watching the next big game! Serve with ranch dressing and stalks of celery for an even more realistic experience. A deep-fat fryer comes in handy here, but a pot filled with 4 inches (10 cm) oil heated to 350°F (180°C) will do just fine.

1 **pound (454 g) tofu, seitan, or tempeh**

FOR BATTER:

1 **cup (125 g) all-purpose flour**
1/2 **teaspoon ground pepper to taste**
1/4 **teaspoon paprika**
1/4 **teaspoon dried oregano**
1/4 **teaspoon dried basil**
1 **cup (235 ml) full-fat coconut milk**
Vegetable oil for frying

FOR BUFFALO SAUCE:

1/2 **cup (112 g) nondairy butter, melted**
2 **to 4 tablespoons (30 to 60 ml) hot sauce to taste**
1 **teaspoon garlic powder**
Salt and pepper to taste

FOR COOL RANCH DIPPING SAUCE:

1 **cup (224 g) vegan mayonnaise, store-bought or homemade (page 66 or 67)**
1/2 **cup (120 g) nondairy sour cream, store-bought or homemade (page 20)**
1/2 **teaspoon dried parsley**
1/2 **teaspoon dried dill**
1/4 **teaspoon garlic powder**
1/4 **teaspoon onion powder**
1/4 **teaspoon ground cumin**
1/8 **teaspoon salt**
1/8 **teaspoon black pepper**

TO MAKE THE WINGS: Cut tofu, seitan, or tempeh into 16 "chicken wing"-size pieces.
Heat oil to 350°F (180°C).
In a shallow dish, combine flour, pepper, paprika, oregano, and basil.
Place coconut milk in a second shallow dish.
Double dredge a "wing" by dipping it into coconut milk, then into flour mixture, then back into milk, and back into flour.
Place wing in hot oil and fry until golden, about 1 minute.
Transfer wing to a paper towel–lined plate to absorb excess oil.
Repeat with remaining wings.

TO MAKE THE BUFFALO SAUCE: Stir hot sauce, garlic powder, salt, and pepper into melted butter.
Place fried wings in a large bowl and top with Buffalo Sauce.
Gently toss to coat.

TO MAKE THE DRESSING: Combine all ingredients until well incorporated.
Store dressing in airtight container for up to 2 weeks in fridge.

YIELD: 16 wings, 1 1/2 cups (350 ml) Cool Ranch Dipping Sauce

CRISPY TOFU VEGGIE SPRING ROLLS

 Meat sub: extra-firm tofu

 Corn Free

Celine used to be addicted to shrimp- and beef-filled spring rolls but finds that this tofu version is just as tasty and definitely healthier for everyone, especially the shrimp and cows. Serve with Tamarind Almond Dip (page 257).

2 tablespoons (30 ml) peanut oil

1 pound (454 g) extra-firm tofu, drained and pressed, cut into ¹/₂-inch (1.27 cm) cubes

¹/₂ teaspoon fine sea salt

¹/₂ teaspoon white pepper

2 tablespoons (30 ml) reduced-sodium soy sauce
   Generous squirt sriracha to taste

¹/₄ cup (40 g) chopped onion

3 cloves garlic, peeled and minced

1 inch (2.5 cm) fresh ginger, peeled and grated

2 tablespoons (30 ml) sesame oil (not toasted)

5 cups (600 g) shredded cabbage

2 cups (390 g) cooked brown jasmine rice

40 spring roll rice paper skins

ADD 1 TABLESPOON (15 ml) of oil to skillet and cook tofu, salt, and pepper over medium-high heat for 10 minutes, until tofu turns golden brown. Transfer to a plate and set aside.

Add soy sauce, sriracha, onion, garlic, and ginger to sesame oil and cook for 2 minutes. Add cabbage and cook over medium-high heat until wilted, about 4 minutes.

Combine with tofu and rice.

Rice paper must be softened before use. Immerse rice paper one sheet at a time in warm water. Soak until soft, approximately 1 minute. Handle carefully as paper breaks easily. Drain on clean kitchen towel before rolling.

Place 1 heaping tablespoon (30 g) of tofu mixture in lower middle of rice paper.

Tightly roll out and fold ends when arriving at the top of the rice paper, rolling out once more to tuck in seams.

Heat remaining tablespoon (15 ml) oil in large skillet. Fill with as many spring rolls as possible without overcrowding, and cook over medium-high heat until golden brown on both sides. Repeat with remaining spring rolls, adding more oil if needed.

YIELD: About 40 rolls

SHEPHERD'S PIE

 Meat sub: extra-firm tofu or tempeh

Comfort food all the way, this hearty dish uses tofu or tempeh to kick meat to the curb.

FOR GRAVY:

- 1 cup (235 ml) vegetable broth
- 1/2 cup (120 ml) sherry or dry white wine
- 2 tablespoons (30 ml) soy sauce
- 1 tablespoon (16 g) mild Dijon mustard
- 2 cloves garlic, peeled and minced
- 1/4 cup (30 g) nutritional yeast
- 2 tablespoons (16 g) cornstarch

FOR FILLING:

- 1 pound (454 g) extra-firm tofu, drained and pressed, or tempeh, cut into 1/2-inch (1.27 cm) cubes
- Salt and pepper to taste
- 1 teaspoon celery seeds
- 6 ounces (170 g) chopped onion

- 1 tablespoon (15 ml) peanut oil
- 1 pound (454 g) frozen mixture of peas, carrots, and corn
- 1 tablespoon (3 g) dried oregano
- 1/4 cup (15 g) chopped fresh parsley
- 1 teaspoon dried rosemary

FOR POTATO TOPPING:

- 6 medium russet potatoes, cut into large chunks, boiled until tender
- 1/4 cup (56 g) nondairy butter, plus extra 2 tablespoons (28 g) to brown top
- 1/2 cup (120 ml) unsweetened nondairy milk
- Salt to taste
- 1 teaspoon red pepper flakes

TO MAKE THE GRAVY: Whisk all ingredients in medium bowl. Set aside.

TO MAKE THE FILLING: Cook tofu or tempeh, salt, pepper, celery seeds, onion and oil in skillet over medium-high heat until golden brown, stirring often, about 10 minutes.

Stir in frozen vegetables, herbs, and gravy; cook over medium heat until thickened, about 3 minutes. Set aside.

TO MAKE THE POTATO TOPPING: While the potatoes are still hot, mash them together with 1/4 cup of butter, milk, and salt, adding more milk if needed to get desired consistency.

Preheat oven to 350°F (180°C, or gas mark 4). Place tofu or tempeh filling in a 9-inch (23 cm) square baking dish. Top evenly with potatoes. Dot remaining butter on top.

Sprinkle with red pepper flakes. Place baking dish on rimmed baking sheet.

Bake for 25 minutes, or until golden brown. For a crisp topping, put dish under broiler for a couple of minutes. Watch carefully so it doesn't burn!

YIELD: 6 servings

BUTTERNUT LENTIL BURGERS

 Meat sub: vital wheat gluten and red lentils

 Corn Free **Soy Free**

A quick-and-easy way to get your protein in the form of gluten, or you can make a gluten-free version by simply taking out the vital wheat gluten. Serve with Smoky Potato Wedges (page 131).

 1 **cup (200 g) cubed butternut squash**
 2 **tablespoons (30 ml) extra-virgin olive oil**
 1/4 **cup plus 1 tablespoon (105 g) prepared tomato chutney**
 1/2 **cup (140 g) cooked red lentils**
 2 **tablespoons (30 ml) sesame oil (not toasted)**
 1 **tablespoon (6 g) garam masala**
 1 **teaspoon fine sea salt**
 1 **cup (80 g) panko or regular bread crumbs**
 1 **cup (144 g) vital wheat gluten**

PREHEAT OVEN to 400°F (200°C, or gas mark 6). Toss cubes with olive oil and spread on baking sheet. Roast for 45 minutes or until tender. Set aside.

Lower temperature to 350°F (180°C, or gas mark 4). Line a cookie sheet with parchment paper or silicone baking mat. Have handy another piece of parchment paper or baking mat, along with another cookie sheet.

Combine squash cubes, chutney, lentils, oil, garam masala, and salt in a large bowl. Combine well, using a potato masher.

Add bread crumbs and gluten; stir well.

Divide preparation into 1/2-cup (120 g) portions. Shape into burgers, about 1/2-inch (1.27 cm) thick. You should have 6 burgers.

Place burgers on prepared cookie sheet. Cover with second layer of parchment paper or baking mat and top with second cookie sheet to keep the burgers from puffing up too much.

Bake for 35 minutes (no need to flip halfway through), or until burgers are golden brown and firm.

YIELD: 6 burgers

BAKED SEITAN CUTLETS

 Meat sub: vital wheat gluten

If you like your seitan only lightly flavored so that it can join in pretty much any dish you have in mind, this one's for you. Crispy and delicious without any further preparation, it also works wonderfully in dish where seitan is commonly used.

Nonstick cooking spray

2 tablespoons (30 ml) tamari

1/4 cup (60 g) ketchup

1/2 cup (120 ml) lukewarm water, up to 1/4 cup (60 ml) more if needed

1 tablespoon (15 ml) peanut oil

11/2 cups (216 g) vital wheat gluten

2 tablespoons (15 g) nutritional yeast

2 tablespoons (24 g) coarse almond meal or (14 g) ground flaxseeds

1/2 teaspoon garlic powder

1 teaspoon granulated onion

1/2 teaspoon all-purpose seasoning, such as McCormick

PREHEAT OVEN to 350°F (180°C, or gas mark 4). Lightly coat a large baking sheet with cooking spray.

Whisk together tamari, ketchup, water, and oil in a small bowl.

Add gluten, yeast, almond meal or flaxseed, garlic powder, onion, and seasoning, stirring with a spoon. Add more water, a little at a time, if the preparation is too dry and use your hands to incorporate. Knead for a minute.

Let rest for 5 minutes.

Cut dough into four equal parts.

Shape pieces into flat, thin cutlets, about 4 inches (10 cm) long, and less than 1/2-inch (1.27 cm) thick

Place cutlets on prepared sheet, leaving room between each. Bake for 20 minutes.

Flip cutlets over and bake for another 20 minutes.

YIELD: 4 cutlets

SMOKY POTATO WEDGES

 Meat sub: liquid smoke (for flavor)

 Corn Free   **Soy Free**   **Wheat Free**

If it's the smokiness and crispiness of bacon that keeps you coming back for more, why not simply add liquid smoke to healthy stuff and call it a day?

Be sure to dry the potatoes when they are cut to allow them to crisp up nicely. Another way to ensure extra crispness is to preheat the baking sheet at the same time the oven heats.

- 2 large russet potatoes, each cut into 8 wedges
- 1¹/₂ tablespoons (23 ml) peanut oil
- 1 tablespoon (15 ml) apple cider vinegar
- 1¹/₂ teaspoons liquid smoke
- ¹/₂ teaspoon fine sea salt to taste
  White or black pepper to taste
- 2 tablespoons (16 g) potato starch

PREHEAT OVEN to 425°F (220°C, or gas mark 7). Place a large rimmed baking sheet in the oven while it preheats.

Toss potato wedges in a large bowl with oil, vinegar, liquid smoke, salt, and pepper. Sprinkle with potato starch; toss again.

Spread wedges on the baking sheet. Bake for 25 minutes; flip wedges over. Bake for another 25 minutes, or until golden brown, crispy, and tender to the fork.

Add more salt and pepper before serving, if desired.

YIELD: 16 wedges

Variations: Be creative and experiment with seasonings such as herbes de Provence, paprika and cayenne pepper, Creole seasoning, or garam masala.

"BEEF" AND BROCCOLI BOWLS

 Meat sub: seitan

 Corn Free   **Quick and Easy**

In order to be true to the "beef" texture, we suggest using prepared seitan for the meat, but you can use tofu or tempeh here, too.

FOR SAUCE:
- 1/4 cup (84 g) brown rice syrup
- 2 tablespoons (30 ml) reduced-sodium tamari or soy sauce
- 2 teaspoons mirin
- 1 tablespoon plus 1 teaspoon (20 ml) sesame oil (not toasted)
- 2 teaspoons arrowroot powder
- 1 teaspoon minced garlic
- 1/2 teaspoon red pepper flakes to taste

FOR BOWLS:
- 2 cups (370 g) cooked brown rice
- 2 cups (312 g) steamed broccoli florets
- 8 ounces (227 g) chopped prepared seitan, store-bought or homemade (page 110)
- 1 green onion, chopped
- 1 teaspoon sesame seeds

TO MAKE THE SAUCE: Whisk all ingredients in a medium saucepot and heat over low heat, stirring often, until heated through and slightly thickened.

TO MAKE THE BOWLS: Divide rice between two bowls. Top with broccoli.
   Add seitan to the sauce, toss until coated and just heated through. Divide seitan between bowls. Sprinkle with chopped onion and sesame seeds.

YIELD: 2 bowls

GREEN TEMPEH VEGGIE FEAST

 Meat sub: tempeh

It's so tasty and filling, being green. Way to show your environmentalist colors by choosing not to eat meat! This dish is delicious on its own or served with rice, mashed potatoes, or pasta.

FOR PESTO:
- 1/3 cup (40 g) roasted salted pepitas (hulled pumpkin seeds)
- 1 cup (40 g) packed basil leaves
- 1 clove garlic, peeled and chopped
- 1/2 teaspoon red pepper flakes
- 1/4 cup (60 ml) extra-virgin olive oil

FOR VEGGIE COMBO:
- 2 tablespoons (30 ml) peanut oil
- 1 pound (454 g) Brussels sprouts, trimmed, and thinly sliced (use food processor for best results)
- 2 cups (178 g) leeks, trimmed and sliced into 1/2-inch (1.27 cm) pieces, thoroughly cleaned
- 2 cloves garlic, peeled and minced
- 1/2 teaspoon ground black pepper to taste
- 1/2 teaspoon fine sea salt to taste
- Dash vegan Worcestershire sauce
- 8 ounces (227 g) tempeh, crumbled into small bits
- 3/4 cup (180 ml) vegetable broth, more if needed

TO MAKE THE PESTO: Place pepitas in food processor. Process until finely ground. Add basil, garlic, and red pepper flakes.

Process until ground. Drizzle in oil, with processor running, until pesto reaches desired consistency. Set aside.

TO MAKE THE VEGGIE COMBO: Heat oil in a large saucepan over medium heat.

Add Brussels sprouts, leeks, garlic, pepper, salt, Worcestershire sauce, and tempeh. Stir and cook until tempeh starts to brown, about 4 minutes.

Add broth and cook until liquid has almost evaporated and the veggies are tender, about 6 minutes. Add more broth if needed.

Stir pesto into veggies. Serve hot.

YIELD: 4 servings

MAMOU'S SPAGHETTI ALLA BOLOGNESE

 Meat sub: firm tofu

The original version contained meat, easily replaced with tofu here. Beef-flavored seitan or ground faux beef would be great, too!

1/2 cup (10 g) dried shiitake or other mushrooms
  2 medium (200 g) russet potatoes, peeled and cut into 1-inch (2.5 cm) cubes
  6 ounces (170 g) whole-grain spaghetti
  5 Roma tomatoes
  1 tablespoon (15 ml) olive oil
  2 cloves garlic, peeled and minced
  1 medium onion, chopped
1 1/2 tablespoons (15 g) chopped shallot
  8 ounces (227 g) firm tofu, crumbled 🥫
  2 cups (300 g) diced green and yellow bell peppers
15 small (150 g) white mushrooms
  2 tablespoons (5 g) chopped fresh basil
  1 tablespoon (3 g) minced fresh thyme
  1 dried bay leaf
  1 teaspoon fine sea salt to taste
    Black pepper to taste
  1 tablespoon (17 g) tomato paste
  1 can (15 ounces, or 425 g) diced tomatoes, with liquid
1/2 cup (58 g) Walnut "Parmesan" Sprinkles (page 51)
1/3 cup (20 g) chopped fresh parsley

SOAK MUSHROOMS in hot water for 10 minutes.

Place potatoes in large saucepan and add enough water to cover by 2 inches (5 cm).

Bring water to a boil; cook for 5 minutes. Add pasta and cook according to package instructions, until potatoes are tender and pasta is al dente.

Drain and set aside.

In the same saucepan, bring water to a boil and blanch tomatoes for 30 seconds. Peel, core, and slice tomatoes.

Drain and rinse rehydrated mushrooms. Slice if they are large.

In large saucepan, heat oil over medium-high. Add garlic, onion, and shallot; cook for 2 minutes.

Add tofu and cook for 2 minutes. Add bell peppers, rehydrated and fresh mushrooms, basil, thyme, bay leaf, salt, and pepper. Cook for 3 minutes, stirring occasionally.

Stir in tomato slices, tomato paste, and diced tomatoes; simmer for 5 minutes.

Stir in potatoes, pasta, Walnut "Parmesan" Sprinkles, and parsley. Simmer until heated through, adding more salt and pepper if desired. Remove bay leaf before serving.

YIELD: 4 to 6 servings

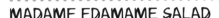

MADAME EDAMAME SALAD

 Meat sub: edamame

 Corn Free **Quick and Easy**

It's good to take a break from mimicking meat every now and then. Here, edamame brings protein to the table that pairs nicely with the refreshing flavors and nutritional punch delivered by the crunchy vegetables.

FOR DRESSING:

1/4 cup (40 g) raisins
3 tablespoons (45 g) stone-ground mustard
2 tablespoons (42 g) agave nectar
2 teaspoons dried thyme
1/4 cup (60 ml) white wine
2 tablespoons (30 ml) red wine vinegar
2 tablespoons (10 g) finely ground walnuts
1/4 cup (40 g) chopped green onions
1 clove garlic, peeled and minced
2 tablespoons (30 ml) extra-virgin olive oil
Freshly ground black pepper to taste
Pinch red pepper flakes to taste
1/2 teaspoon fine sea salt to taste

FOR SALAD:

1/4 cup (15 g) chopped fresh parsley
1 2/3 cups (255 g) ready-to-eat shelled edamame
2 endives, thinly sliced
1 small head radicchio, chopped
10 radishes, thinly sliced

TO MAKE THE DRESSING: Combine all ingredients in a medium saucepan. Bring to a low boil. Simmer for 4 minutes to let flavors develop. Enjoy warm or cold.

TO MAKE THE SALAD: In a large bowl, combine veggies and enough dressing to lightly coat.

YIELD: 4 servings

ALL-AMERICAN HOT DOGS

 Meat sub: vital wheat gluten and tofu for pork or beef hot dogs

Whether served in a bun with ketchup and mustard, or topped with Chili Con Carne (page 118), onions, and a healthy dose of nondairy shredded cheese, or made into Beer Battered Corndogs (right page), these everyday wieners are sure to please everyone—even the kids!

2 cups (288 g) vital wheat gluten flour

1 cup (120 g) whole wheat pastry flour

2 tablespoons (14 g) smoked paprika

1 teaspoon garlic powder

1 teaspoon onion powder

1 teaspoon maca powder (optional)

$^1/_2$ teaspoon turmeric

$1^1/_2$ (355 ml) cups water

4 ounces (112 g) extra-firm tofu, drained and pressed

$^1/_2$ cup (120 ml) canola or vegetable oil

$^1/_4$ cup (60 ml) soy sauce or tamari

$^1/_4$ cup (84 g) brown rice syrup

2 tablespoons (33 g) tomato paste

1 tablespoon (15 ml) liquid smoke

PREHEAT OVEN to 350°F (180°C, or gas mark 4).

Mix together flours, paprika, garlic powder, onion powder, maca powder, and turmeric.

In a blender, purée water, tofu, oil, soy sauce or tamari, rice syrup, tomato paste, and liquid smoke.

Add wet ingredients to dry and mix.

Divide dough into 8 to 12 pieces, depending on how large you like your weiners.

Tear off 8 to 12 pieces of aluminum foil, each about 6 x 12 inches (15 x 30 cm).

Form each piece of dough into a sausage shape and place near the long edge of the foil. Roll up the foil and twist the ends tight.

Place seam side down on a baking sheet and bake for 30 to 40 minutes, or until firm.

Remove from oven and let sit until cool enough to handle before unwrapping.

YIELD: 8 to 12 hot dogs

BEER BATTERED CORN DOGS

 Meat sub: vegan hot dogs for hot dogs

Who doesn't love a corn dog? Made with beer, this batter cooks up quickly, is crispier than the traditional carnival fare, and a tad more sophisticated. We think whole-grain mustard is the perfect condiment for this spiffy dog.

- 3/4 cup (93 g) all-purpose flour
- 3/4 cup (105 g) yellow cornmeal
- 1 cup (235 ml) vegan beer
- 2 tablespoons (30 ml) vegetable oil
- 1 tablespoon (21 g) agave nectar
- 1 teaspoon Dijon mustard
- 1/2 teaspoon baking soda
- 1/2 teaspoon baking powder
- 1/4 teaspoon paprika
- 6 vegan hot dogs, store-bought or homemade (page 136)
- 6 Popsicle sticks, skewers, or disposable chopsticks
  Vegetable oil for frying

IF YOU DON'T HAVE A DEEP FRYER, fill a pot with 4 inches (10 cm) oil. Preheat oil to 350°F (180°C, or gas mark 4). Line a plate or tray with paper towels.

Mix flour, cornmeal, beer, 2 tablespoons vegetable oil, agave, mustard, baking soda, baking powder, and paprika until smooth. Transfer batter to a tall glass, like a pint glass, to make dipping easy.

Insert a Popsicle stick, skewer, or disposable chopstick into the bottom of each hot dog, making sure to leave enough stick showing for a handle. Dip each hot dog in batter to coat.

Wear a kitchen glove to protect your hands from hot oil splatter.

Submerge the battered dog, using the handle, into the hot oil to fry until golden and crispy, about 1 minute.

Transfer corn dogs to paper towel–lined plate to drain excess oil.

YIELD: 6 full-size corn dogs or
18 corn-dog nuggets

Note: You can omit the sticks and make corn dog nuggets by cutting hot dogs into 2- or 3-inch (5 or 8 cm) pieces and using a slotted spoon to remove them from the oil.

PIZZA PEPPERONI TOPPING

 Meat sub: TVP granules and vital wheat gluten

We offer two versions of what you'll discover is an unbelievably realistic rendition of the real thing.

FOR GLUTEN-FREE VERSION:
- 1 cup (96 g) TVP granules
- 1/2 teaspoon ground pepper
- 1 tablespoon (7 g) paprika
- 1 tablespoon (8 g) garlic powder
- 1 teaspoon aniseed
- 1 teaspoon sea salt
- 1/2 to 1 teaspoon red pepper flakes to taste
- 1 teaspoon Sucanat
- 1 teaspoon dried basil
- 1/2 to 1 teaspoon cayenne pepper to taste
- 2 tablespoons (30 ml) olive oil
- 1–2 tablespoons (15–30 ml) liquid smoke, to taste
- 1 cup (235 ml) water

FOR SOY-FREE VERSION:
- 1 cup (144 g) vital wheat gluten
- 1/2 teaspoon ground pepper
- 1 tablespoon (7 g) paprika
- 1 tablespoon (8 g) garlic powder
- 1 teaspoon anise seed
- 1 teaspoon sea salt
- 1/2 to 1 teaspoon red pepper flakes to taste
- 1 teaspoon Sucanat
- 1 teaspoon dried basil
- 1/2 to 1 teaspoon cayenne pepper to taste
- 2 tablespoons (30 ml) olive oil
- 1–2 tablespoons (15–30 ml) liquid smoke, to taste
- 1/4 cup (60 ml) water, more if needed

TO MAKE GLUTEN-FREE VERSION: In a microwave-safe bowl, combine first TVP with herbs and spices, through cayenne pepper.

Stir in oil, liquid smoke, and water. Cover tightly with plastic wrap and microwave on high for 6 minutes. Carefully remove from microwave, stir, and let cool.

Alternatively, combine TVP with herbs and spices, through cayenne pepper. Bring water to a boil, add in oil and liquid smoke, stir into TVP and spice mixture, cover, and let sit for 10 minutes. Fluff with a fork.

TO MAKE SOY-FREE VERSION: Combine first 10 ingredients (through cayenne pepper) in medium bowl.

Place oil and liquid smoke in skillet. Combine with dry ingredients, using a spatula. Add water and mix it all in, using your fingertips, crumbling the mixture until no dry spots are left. Add more water if mixture is too dry.

Fry crumbles over medium heat, breaking apart larger crumbles, and stirring often for about 10 minutes, or until browned and cooked through.

YIELD: 2 1/2 cups (312 g), enough to top 1 large pizza

BABY BACK RIBS

 Meat sub: vital wheat gluten

 For the more experienced cook

We know that most vegans are pretty turned off by the idea of gnawing meat off bones, but we also know that there are a lot of vegans who used to eat meat and remember that it tasted good. We thought it would be fun to offer a few recipes that truly mimic the look, taste, texture, and smell of some of those remembered dishes.

- 3 cups (432 g) vital wheat gluten flour
- 1 cup (120 g) whole wheat flour
- 1 tablespoon (8 g) garlic powder
- 1 tablespoon (8 g) onion powder
- 1 teaspoon black pepper
- 1 teaspoon paprika
- 1 teaspoon chipotle powder
- 1/3 cup (80 ml) olive oil
- 1/3 cup (80 ml) Bragg Liquid Aminos, tamari, or soy sauce
- 1/3 cup (80 ml) BBQ sauce
- 1 tablespoon (15 ml) liquid smoke
- 1 cup (235 ml) vegetable broth
- 6 bones, optional (see note)
- 1 cup (235 ml) extra BBQ sauce, for baking and basting

Note: Disposable bamboo chopsticks or even Popsicle sticks work wonders here. but if you want to kick it up a notch. try food-grade cedar cut into 1/2 x 6-inch (1.27 x 15 cm) "bones," making sure to sand down any rough edges. Whichever you choose. start by soaking them in a mixture of 2 cups (470 ml) warm water mixed with 2 tablespoons (30 ml) liquid smoke to add some smoky flavor to the ribs.

PREHEAT OVEN to 350°F (180°C, or gas mark 4).

Line a baking pan with foil or parchment paper. In a large mixing bowl, mix together dry ingredients. In a separate bowl, mix together wet ingredients.

Add wet ingredients to the dry and knead for 5 minutes. Let mixture rest for 20 minutes.

Shape mixture into a large rectangle, about 5 x 10 inches (13 x 25 cm). If using bones, push them through the mixture, at equal distances from each other.

Pour 1/2 cup (120 ml) of barbecue sauce into the bottom of the baking pan.

Place "rack of ribs" in the pan.

Spread the other 1/2 cup (120 ml) barbecue sauce over the top.

Cover pan tightly with foil and bake for 90 minutes.

You can enjoy these as is, or refrigerate for later use. To reheat, fire up the grill to give ribs extra smoky flavor and grill marks. Brush with even more barbecue sauce while grilling!

YIELD: 6 large ribs

BLACK FOREST BACON

 Meat sub: vital wheat gluten

 For the more experienced cook

Prepare yourself for sweet-smoky-peppery goodness. And when we say smoky, we aren't exaggerating: If you aren't careful, your house will be full of smoke when preparing this, so pay close attention!

FOR BACON:

- 2 cups (288 g) vital wheat gluten flour
- 1/2 cup (60 g) whole wheat pastry flour
- 1 tablespoon (7 g) smoked paprika
- 1 tablespoon (8 g) garlic powder
- 1 teaspoon ground black pepper
- 1/4 cup (60 ml) canola or other mild-flavored vegetable oil
- 1/4 cup (60 ml) maple syrup
- 1 cup (235 ml) vegetable broth or water
- 2 tablespoons (30 ml) liquid smoke

FOR GLAZE:

- 1/4 cup (60 ml) maple syrup
- 1/4 cup (55 g) firmly packed brown sugar
- 3 tablespoons (18 g) ground black pepper
- 2 tablespoons (30 ml) liquid smoke
- 2 teaspoons sea salt
- Aluminum foil
- Nonstick cooking spray
- Additional oil for frying

TO MAKE THE BACON: Preheat oven to 350°F (180°C, or gas mark 4).

In a mixing bowl, combine gluten flour, pastry flour, paprika, garlic powder, and pepper.

Combine oil, syrup, broth or water, and liquid smoke in a separate bowl.

Add wet mixture to dry and knead together until dough forms. If it is too dry, add more broth or water. Set aside.

TO MAKE THE GLAZE: Whisk all ingredients in a small bowl.

Place a large piece of foil on a rimmed baking sheet. Spray foil with nonstick spray.

Place the dough in the center of the foil and shape into a rectangular loaf about 8 x 6 x 1 inches (20 x 15 x 2.5 cm).

Reserving about 1/4 cup (60 ml) of glaze, carefully pour the rest over the loaf.

Carefully fold the foil around the loaf and seal completely. This is an important step. You don't want the glaze to leak out of the foil onto the pan. It will burn and cause a smoky mess!

Bake for 45 minutes.

Remove from oven and let cool for a few minutes. Increase heat to 450°F (230°C, or gas mark 8).

Carefully open the top of foil envelope, pour in reserved 1/4 cup (60 ml) glaze, and return to oven for 10 more minutes.

Watch carefully as you don't want the glaze to leak out.

Remove baking sheet from oven and let cool completely. Or refrigerate until ready to slice and fry.

Using a sharp knife, cut thin "bacon" strips from the slab, and fry in oil in skillet set over high heat until crispy.

YIELD: About 20 slices

IMITATION BACON BITS

 Meat sub: TVP for bacon bits

 Gluten Free

These are so easy to whip up, you can have bacony goodness whenever you want! Sprinkle on top of baked potatoes, in tofu scrambles, or anywhere you would use bacon bits.

2 **tablespoons (30 ml) liquid smoke**
1 **scant cup (205 ml) water**
1 **cup (96 g) TVP granules**
1/4 **teaspoon salt**
   Few drops beet juice for red food coloring, optional
3 **tablespoons (45 ml) canola or other vegetable oil**

ADD LIQUID SMOKE to a 1-cup (235 ml) measuring cup; add water to fill to 1 cup.

In a microwave-safe dish, combine liquid smoke mixture, TVP, salt, and coloring, if using. Cover tightly with plastic wrap and microwave on high for 5 minutes.

(Alternatively you can bring water to a boil, pour over TVP, stir in liquid smoke and salt, cover and let stand for 10 minutes.)

Carefully remove plastic wrap.

Preheat a skillet with oil over medium-high heat.

Add reconstituted TVP to the pan and toss to make sure it gets coated with oil. Fry until desired crispness is reached, stirring often. You don't necessarily want to brown the bits, but rather dry them out, which should take about 10 minutes.

Let cool completely before transferring to an airtight container. Store in the fridge. Keeps well for at least a week.

YIELD: 1 cup (65 g)

ORANGE COUNTY ROLLS

 Fish sub: just naturally fish free!

This fun, easy roll takes the fear and fuss out of sushi. Serve with Spicy Sushi Sauce for dipping and enjoy your sushi the way the Real Housewives do.

FOR ROLLS:

- 2 cups (360 g) uncooked sushi rice
- 3 cups (705 ml) water
- 2 tablespoons (42 g) agave nectar
- 2 tablespoons (30 ml) rice wine vinegar
- Salt to taste
- 4 toasted nori sheets
- 1/2 cup (120 g) nondairy cream cheese
- 1 carrot, grated
- 1 bell pepper, cored, seeded and sliced julienne style
- 1/2 medium red onion, sliced julienne style
- 1 cucumber, seeded and cut into matchsticks
- 1 ripe avocado, sliced

FOR SPICY SUSHI SAUCE:

- 1/2 cup (112 g) vegan mayonnaise, store-bought or homemade (page 66 or 67)
- 2 tablespoons (30 ml) sriracha
- 1/4 teaspoon toasted sesame oil

TO MAKE THE ROLLS: Add rice, water, agave, vinegar, and salt to a rice cooker. Or prepare rice according to package instructions adding agave, vinegar, and salt to the water.

While rice is cooking, prep vegetables. Let rice cool before assembling your rolls.

Place 1 sheet of nori, shiny side down, on sushi mat or on a very dry surface.

Add one-quarter of the rice to the long edge of the nori, spreading it evenly, so that it covers one-third of the sheet. Spread 2 tablespoons (30 g) cream cheese along the center of the rice.

Follow with a layer of carrot, a layer of bell pepper, a layer of onion, a layer of cucumber, and finally a layer of avocado.

Carefully and tightly roll the sushi like a burrito. The tighter you roll it, the easier it will be to cut.

Repeat with remaining 3 sheets of nori. Using a sharp knife, cut each roll into 8 equal pieces.

TO MAKE THE SAUCE: Whisk together mayo, sriracha, and sesame oil.

YIELD: 32 pieces and ½ cup (120 ml) sauce.

FISH-Y STICKS WITH TARTAR SAUCE

 Fish sub: nuts and seeds for texture, seaweed for fishy flavor

Treat your toddlers (and yourself!) to these tasty and healthful little fish-y fingers.

FOR FISH-Y STICKS:

- 2 cups (470 ml) vegetable broth
- 1 cup (168 g) dry quinoa
- 1 cup (96 g) TVP granules
- ¼ cup (35 g) pumpkin seeds
- ¼ cup (32 g) sunflower seeds
- 2 tablespoons (16 g) sesame seeds
- ¼ cup (29 g) wheat germ
- ½ cup (56 g) raw cashews, ground into a powder
- ½ cup (72 g) vital wheat gluten flour
- ¼ cup (26 g) flaxseed meal
- ¼ cup (60 ml) olive oil
- ¼ cup (30 g) dulse or kelp granules
- 2 tablespoons (30 g) minced garlic
- 1 teaspoon paprika
  Salt and pepper to taste
- 1 cup (235 ml) full-fat coconut milk
- 1 cup (80 g) panko bread crumbs
- ½ cup (120 ml) vegetable oil for frying

FOR TARTAR SAUCE:

- 1 cup (224 g) vegan mayonnaise, store-bought or homemade (page 66 or 67)
- 2 tablespoons (30 g) sweet pickle relish
- 1 tablespoon (7 g) dried minced onion
- 2 tablespoons (30 ml) fresh lemon juice
- 1 teaspoon dried or 1 tablespoon (1 g) fresh dill
  Salt and pepper to taste

Note: If you cannot find dulse or kelp granules, you can make your own by grinding sheets of kelp or dulse into granules in a food processor.

TO MAKE THE FISH-Y STICKS: Bring broth to a boil.

Add quinoa and reduce to a simmer. Allow quinoa to absorb most of the liquid, but leave it soupy.

Stir in TVP, cover, remove from heat, and let stand for at least 10 minutes.

In a dry skillet, add pumpkin seeds, sunflower seeds, sesame seeds, and wheat germ. Toast over medium heat, stirring constantly, being careful not to burn.

Add toasted seeds, ground cashews, gluten flour, flaxseed meal, olive oil, dulse granules, garlic, paprika, salt, and pepper to the TVP and quinoa mixture. Mix thoroughly, using your hands.

Form into fish stick-size fingers, and refrigerate for 1 hour to stiffen up. Place coconut milk in a shallow bowl, place bread crumbs in a shallow dish or plate, and preheat skillet with oil over medium-high heat.

Dip a fish stick in coconut milk, coat in bread crumbs, and place in the skillet.

Fry for 3 to 5 minutes per side, until golden brown.

TO MAKE THE TARTAR SAUCE: Whisk together all ingredients and store in an airtight container in the refrigerator until ready to use.

YIELD: 24 fish sticks and 1 cup (250 g) tartar sauce

HAPPY SEA TEMPEH SALAD

 Fish sub: tempeh for texture, hijiki for fishy flavor

 Corn Free   **Gluten Free**

 Quick and Easy   **Wheat Free**

This no-cook salad comes together in mere minutes and tastes wonderful as a sandwich, a melt, on crackers, or over greens. The seaweed used in this recipe gives it just a hint of that fish-y flavor one looks for in a mock-tuna salad.

¼ cup (20 g) hijiki, reconstituted
   in ½ cup (120 ml) warm water
8 ounces (227 g) tempeh
1 stalk celery, chopped
½ cup (80 g) diced red onion
¼ cup (32 g) hulled sunflower seeds
2 tablespoons (30 g) pickle relish
1 tablespoon (1 g) fresh dill or 1 teaspoon dried dill
¼ to ½ cup (56 to 112 g) vegan mayonnaise to taste
   Salt and pepper to taste

ADD HIJIKI TO WARM water and let sit while prepping the rest of the salad.

Drain excess liquid before adding to the salad.

In a mixing bowl, crumble tempeh until it almost broken into individual beans. Add remaining ingredients and mix well.

Chill overnight to let flavors meld.

YIELD: 2 cups (450 g) or enough for about 6 sandwiches

Note: If you can't find or don't like hijiki, you can use any dried seaweed, even nori sheets, cut into tiny pieces. If you do use nori, you don't need to reconstitute it, just mix it in with the rest of the ingredients.

Some people prefer to simmer or steam their tempeh before using it. This can be a helpful tip for tempeh first-timers, as it tends to mellow the fermented flavor a bit. If you're interested, just steam or simmer your tempeh for about 20 minutes, and let cool completely before crumbling.

SURF AND TURF SALAD

 Fish sub: hijiki or other dried seaweed

 Corn Free **Wheat Free**

A little bit from the sea and a little bit from the earth. This nutty salad is easy to throw together and adds a lot more to a potluck than the traditional pasta salad or coleslaw.

FOR SALAD:

- 2 cups (370 g) cooked cold red quinoa
- 1 cup (280 g) cooked cold green or yellow lentils
- 1 cup (140 g) whole raw cashews
- 1/4 cup (20 g) hijiki or other dried seaweed, reconstituted in 1/2 cup (120 ml) warm water
- 1 red or green bell pepper, cored, seeded, and diced

FOR DRESSING:

- 1/4 cup (60 ml) olive oil
- 2 tablespoons (30 ml) white rice vinegar
- 1 tablespoon (15 ml) soy sauce or tamari
- 1/4 cup (84 g) agave nectar
- 1 teaspoon red pepper flakes to taste
- 1/2 teaspoon ground ginger
- Salt and pepper to taste

DRAIN EXCESS LIQUID from hijiki.
In a medium bowl, toss together salad ingredients.
In a small bowl, whisk together dressing ingredients.
Toss dressing with salad to coat.
Refrigerate for several hours before enjoying for the flavors to develop.

YIELD: 4 to 6 servings

LEAVE THE BEES TO BUZZ!

FOOLPROOF SUBSTITUTIONS FOR ANIMAL BY-PRODUCTS

Chapter 5

From Syrupy Glazes to Jiggling Jellies:
HOW TO SUBSTITUTE FOR HONEY AND GELATIN

ADOPTING A VEGAN LIFESTYLE means more than simply cutting meat and dairy out of your diet. It's also important to keep a close eye on the less obvious manufacturing process and hidden animal by-products that go into anything you decide to support with your hard-earned cash, be it food, cosmetics, or clothing.

Consider the Facts

An animal by-product is anything that is forcefully taken from an animal, when it's still alive or already butchered. This encompasses the wool of a sheep, the honey from bees, the acid boiled out of dried insects to dye foods, the charred animal bones used to whiten the sugar you like to add to your morning coffee. Not very appetizing . . .

These by-products aren't used because they yield better results or are healthier, which they surely aren't, but simply because they are the cheapest alternative. It has been proven that plant-based, cruelty-free methods can be used to yield the same if not superior results.

Here is a list of some of the most common by-products that sneak into the foods we eat and the cosmetics we use. If you ever find yourself in doubt whether an ingredient is plant or animal derived, we strongly recommend contacting the manufacturer. We also recommend you consult *Animal Ingredients A to Z* by EG Smith Collective, for a thorough list of animal-derived ingredients.

- Albumin
- Calcium stearate
- Carmine
- Cochineal
- Capric acid
- Casein
- Clarifying agent, or fining agent
- Gelatin
- Glycerides (mono-, di-, and tri-)

- Isinglass (except Japanese isinglass, which is made from agar)
- Lactose
- Lactic acid (except if made from synthetic lactonitrile)
- Lanolin
- Lard
- Oleic or oleinic acid
- Pancreatin
- Pepsin

- Propolis
- Rennet
- Royal jelly
- Stearic acid or octadecanoic acid
- Suet
- Tallow
- Vitamin D_3
- Whey

Animal By-Products Substitution Guidelines

For more specific guidelines, let us turn to the following chart and then see if we can apply our knowledge to a nonvegan recipe.

IF THE ORIGINAL RECIPE CALLS FOR...	REPLACE WITH...
1 cup (224 g) butter or lard	• 1 cup (224 g) nondairy butter, solid coconut oil, or vegetable shortening
Food dye	• Beet, tomato, spinach, carrot, or blueberry juice or powder. • Store-bought natural food dye, such as India Tree
Gelatin	• Agar, fruit pectin, carrageenan, locust bean gum
1 cup (235 ml) honey	• 1 cup (235 ml) agave nectar, barley malt, pure maple syrup, blackstrap or regular molasses

》 BBQ Beans (page 156)

VEGANIZED!: SAMPLE RECIPE

Take a look at the following traditional recipe for an example of how we would replace the animal by-products and other nonvegan ingredients:

RED WALDORF CAKE

This traditional favorite usually calls for artificial food color. It was adapted from the *Better Homes and Gardens New Cook Book*.

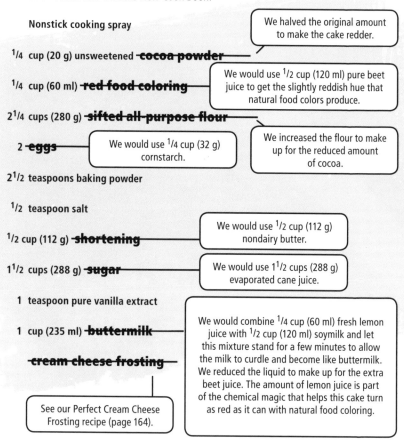

Nonstick cooking spray

We halved the original amount to make the cake redder.

1/4 cup (20 g) unsweetened ~~cocoa powder~~

1/4 cup (60 ml) ~~red food coloring~~

We would use 1/2 cup (120 ml) pure beet juice to get the slightly reddish hue that natural food colors produce.

2 1/4 cups (280 g) ~~sifted all-purpose flour~~

2 ~~eggs~~

We would use 1/4 cup (32 g) cornstarch.

We increased the flour to make up for the reduced amount of cocoa.

2 1/2 teaspoons baking powder

1/2 teaspoon salt

1/2 cup (112 g) ~~shortening~~

We would use 1/2 cup (112 g) nondairy butter.

1 1/2 cups (288 g) ~~sugar~~

We would use 1 1/2 cups (288 g) evaporated cane juice.

1 teaspoon pure vanilla extract

1 cup (235 ml) ~~buttermilk~~

We would combine 1/4 cup (60 ml) fresh lemon juice with 1/2 cup (120 ml) soymilk and let this mixture stand for a few minutes to allow the milk to curdle and become like buttermilk. We reduced the liquid to make up for the extra beet juice. The amount of lemon juice is part of the chemical magic that helps this cake turn as red as it can with natural food coloring.

~~cream cheese frosting~~

See our Perfect Cream Cheese Frosting recipe (page 164).

LIGHTLY COAT two 8-inch (20 cm) round cake pans with cooking spray. In a small bowl, combine cocoa and beet juice. In another bowl, combine flour, cornstarch, baking powder, and salt.

Preheat oven to 350°F (180°C, or gas mark 4). Beat butter with an electric mixer for 30 seconds. Add evaporated cane juice and vanilla; beat until well combined. Beat in cocoa mixture. Alternately add flour mixture and buttermilk, beating on low to medium speed after each addition until just combined. Place batter into prepared pans.

Bake for 35 minutes, or until a toothpick inserted in the center comes out clean. Remove pans and cool for 10 minutes. Remove cake from pans, and cool on wire rack. When cakes have cooled completely, apply frosting.

YIELD: 12 servings

Finding Animal By-Product Substitutes at the Store

Most of the following ingredients can be found at your local health-food store or international market. Some hard-to-find ingredients can be purchased online.

Agar: This is the thickening agent we use the most when looking for a gelatin-like consistency. Used almost exactly like gelatin, this colorless, odorless, and tasteless sea vegetable can be purchased in flakes or powder. Try to find it in its natural form, which looks almost like a long rod of Styrofoam, as this form of agar is much less expensive than the others. If you can find it this way, break off a chunk and grind it into flakes or powder using a very dry food processor or blender. Use 3 tablespoons (15 g) agar flakes or 1^1/$_2$ teaspoons agar powder to thicken 2 cups (470 ml) of liquid. If using an acidic liquid, use a little more agar. Combine with liquid, bring to a boil, and simmer 15 minutes, or until the agar flakes or powder dissolve. Let cool to thicken.

Agave and **other sweeteners:** Liquid sweeteners work perfectly in place of honey, usually at a 1:1 ratio. See chapter 8 (page 226) for more info on these substitutes.

All-natural food colors: Look for brands such as India Tree for cake and cookie decorations and for all-natural plant-based food dyes. Using dehydrated vegetable powders, such as spinach, carrot, tomato, and beet, also works well for adding all-natural color without introducing extra liquid.

Beet powder: Ground from the dried beetroot. The issue with using beet powder in baked goods, namely in Red Velvet cake, is that its pigment reacts to heat and pH, so it won't show its true colors when combined with baking powder and baking soda. Using juice is best for stronger colors in baked or cooked foods. Keep in mind that liquid natural coloring will alter the texture of frosting, so have extra powdered sugar on hand when playing around with coloring.

Carrageenan: A purple seaweed used as a thickener, found in health-food stores. One ounce (28 g) carrageenan thickens 1 cup (235 ml) of liquid. Rinse carrageenan and soak until it swells. Combine with liquid and bring to a boil for 10 minutes. Strain and discard carrageenan.

Fruit pectin: Used to make jams and jellies gel, this handy powder is also used in place of gelatin in many recipes, and can be found in the canning section at the supermarket. Follow instructions on packaging for best results.

Locust bean gum: You may know it as carob gum, since it is extracted from the seeds of the carob tree. Most often used as a thickener in prepackaged foods.

Marshmallows: Sweet and Sara and Dandies are the only two vegan brands we know of as of printing. Both are scrumptious and available in stores and online.

Nondairy butter: It is now possible to purchase animal-friendly and heart-healthy buttery spreads that please the palate just as much as cholesterol-laden dairy butter. See chapter 1 (page 16) for more info on this ingredient.

Vegetable gelatin: Natural Desserts makes both flavored and unflavored vegan gelatin mixes that act and look just like Jell-O.

Making Animal By-Products Substitutes at Home

The most important thing to remember is to check your ingredients when buying prepackaged items. As you grow more accustomed to your vegan kitchen, you will find yourself doing a lot more cooking from scratch and therefore eliminating the possibility of having animal by-products in your meals.

One easy substitute you can make at home is food coloring. If you have access to a vegetable juicer, you can create a plethora of all-natural food dyes right in your own kitchen. Just one half of an average-size red beet will yield enough juice to give a vibrant reddish-pink hue to frostings, cookies, and cakes!

Choose between carrots, spinach, and blueberries to add a whole palette of colors to your goodies. Turmeric does a fantastic job of adding a bright yellow hue to foods. Have fun experimenting!

Recipes without Animal By-Products!

Fasten your seat belts and get ready to enter a wonderful world that is free of gelatin, honey, and other animal products!

. . . .

SUPER SIMPLE CREAMY DIJON DRESSING

 By-product sub: agave for honey

 Corn Free   **Gluten Free**   **Quick and Easy**   **Wheat Free**

We believe this rivals any store-bought honey mustard dressing available, even the vegan varieties. And did we mention that this delectable dressing takes less than 2 minutes to make?

10 ounces (283 g) silken tofu
1/4 cup (84 g) agave nectar
1/4 cup (60 g) mild Dijon mustard
2 tablespoons (30 ml) apple cider vinegar
Salt and pepper to taste

PLACE ALL INGREDIENTS in a blender and purée until silky smooth.

Store in an airtight container in the refrigerator for up to 2 weeks.

YIELD: 2 cups (470 ml)

BBQ BEANS

 By-product sub: agave nectar for honey

 Corn Free   Gluten Free   Low Fat   Nut Free

There's no need for sugar or honey in your BBQ sauce! Agave nectar does the trick beautifully and gives these protein-rich little jewels a sweet and tangy punch.

²/3 cup (160 ml) water

6 ounces (170 g) tomato paste or organic ketchup (for a sweeter sauce)

¹/4 cup (84 g) light agave nectar

2 tablespoons (30 ml) balsamic vinegar

2 tablespoons (30 ml) vegan Worcestershire sauce

2 tablespoons (30 ml) adobo sauce, optional

1 tablespoon (22 g) regular molasses

¹/4 teaspoon cayenne pepper to taste

1 teaspoon liquid smoke, optional

¹/3 cup (53 g) chopped onion

1 large clove garlic, peeled and minced

2 teaspoons whole-grain Dijon mustard

2 to 3 cans (each 15 ounces, or 425 g) beans of choice, drained and rinsed

USING A BLENDER, purée all ingredients (except beans) until smooth. Pour sauce into a large pot. Combine with beans, using the amount that yields the right balance of beans and sauce.

Bring to a low boil, lower heat, and simmer uncovered for 20 minutes, stirring occasionally.

YIELD: 6 side-dish servings

SOUP OF THE SHRIER

 By-product sub: agave for honey

 Gluten Free   **Nut Free**

 Soy Free   **Wheat Free**

This soup is inspired by some of Joni's besties, the Shriers: Laurence, Jennifer, and Baby Berlin. This soup is for them!

- 2 **tablespoons (30 ml) olive oil**
- 1 **leek, whites and light greens only, sliced into thin rings**
- 4 **cloves garlic, peeled and chopped**
- 6 **cups (1.41 L) vegetable broth**
- 3 **ounces (84 g) kale, stemmed, torn into small pieces**
- 1 **can (15 ounce, or 425 g) garbanzo beans or white beans, drained and rinsed**
- 1 **teaspoon red pepper flakes to taste**
- 2 **tablespoons (42 g) agave nectar**
- 1/4 **cup (60 ml) lemon juice**
  Salt and pepper to taste

HEAT OIL in a soup pot.
 Sauté leek and garlic until fragrant and just beginning to brown.
 Add broth.
 Bring to a boil, reduce to a simmer. Add kale, beans, red pepper flakes, and agave.
 Simmer for 15 minutes.
 Remove from heat and stir in lemon juice, salt, and pepper.

YIELD: 4 servings

CHERRY-GLAZED TOFU

 By-product sub: just naturally by-product free!

 Corn Free   **Gluten Free**

 Nut Free   **Wheat Free**

This glaze works well on tofu and tempeh, so pick your favorite protein!

- 1 **cup (235 ml) red wine or pomegranate juice**
- 1/2 **cup (80 g) dried cherries**
- 1/4 **cup (80 g) cherry or raspberry preserves**
- 2 **tablespoons (30 ml) olive oil**
- 1 **tablespoon (15 g) mild Dijon mustard**
- 1 **teaspoon dried thyme or 1 tablespoon (3 g) fresh thyme**
- 1/2 **teaspoon ground black pepper**
- 1/4 **teaspoon salt**
  12 to 16 ounces (336 to 454 g) extra-firm tofu, drained and pressed

PREHEAT OVEN to 350°F (180°C, or gas mark 4).
 Combine all ingredients, except tofu, in a pot and bring to a boil.
 Lower temperature and simmer for 10 to 12 minutes, or until mixture is syrupy and cherries are plump.
 While glaze is simmering, cut tofu into 4 equal "steaks."
 Place tofu steaks in a baking dish that is not too large but big enough for the steaks to have at least 1 inch (23 cm) separating them on all sides.
 Pour reduced glaze over tofu steaks and bake uncovered for 30 minutes.

YIELD: 4 servings

RAW DULSE CRISPS

 By-product sub: just naturally by-product free!

 Corn Free Gluten Free Raw

 Soy Free Wheat Free

One thing about taking on a raw-food lifestyle is the amount of time it takes to prepare foods, unless you plan on eating garden salads and avocados for the rest of your life. This recipe is no exception. However, if you prepare this recipe together with Raw Banana Apple "Leather" (page 206) and Raw Flax and Hemp Seed Crackers (page 176), you'll cut your prep time by 30 percent! A dehydrator makes things easy here, but if you don't have one, an oven set to its lowest temperature will also work.

1 ounce (28 g) air-dried dulse
1 tablespoon (15 ml) cold-pressed extra-virgin olive oil
1 tablespoon (15 g) minced garlic
2 tablespoons (13 g) flaxseed meal
4 peppadew or piquante peppers, seeded and cored
1/2 teaspoon sea salt
1/4 cup (32 g) raw hulled sunflower seeds
2 tablespoons (20 g) flaxseeds

ADD DULSE, oil, garlic, flaxseed meal, peppers, and salt to a food processor and pulse until finely chopped.

Stir in sunflower and flaxseeds.

Spread mixture thinly onto the liner tray of your dehydrator. If you do not have a liner tray, use parchment paper cut to the shape of your trays.

Set dehydrator to 115°F (46°C) and "cook" for 8 hours, or until crispy.

If you do not have a dehydrator, set your oven to the lowest setting.

Leave the oven door cracked open. Try not to let the oven get above 115°F (46°C). Spread mixture thinly on a baking sheet lined with parchment paper or a silicone mat.

Once crispy, use a sharp knife or your hands to break into shards.

YIELD: 12 crackers

CRISPY CHOCOLATY CEREAL CANDY

 By-product sub: nondairy butter for butter

 Corn Free   Quick and Easy

You may remember Puppy Chow from holiday parties and Christmas gift baskets. It's ridiculously easy and quick to prepare! It is perfect for gift giving, as one batch makes enough for about 5 goodie bags.

This can easily be made gluten free by choosing certified gluten-free ingredients.

- 1 cup (176 g) semisweet chocolate chips
- 1/2 cup (128 g) creamy no-stir peanut butter
- 1/2 cup (112 g) nondairy butter
- 1 tablespoon (15 ml) pure vanilla extract
- 1 box (13 ounces, or 369 g) crispy rice pocket cereal, such as Rice Chex
- 3 cups (360 g) powdered sugar

COMBINE CHOCOLATE CHIPS, peanut butter, and butter in a microwave-safe bowl. Microwave on high for 1 minute. Stir mixture. If not completely melted, continue to heat in microwave for 20-second intervals. Alternatively, you can use a double boiler. Stir in vanilla.

Pour cereal into your largest mixing bowl.

Add melted peanut butter and chocolate mixture and toss until cereal is completely coated.

Add powdered sugar 1 cup (120 g) at a time and toss to coat. Let cool completely before packaging.

YIELD: 12 cups (624 g)

SWEET TIROPITAKIA

 By-product sub: tofu for cream cheese, cashews for creaminess, agave for honey

Traditional Greek *tiropitakia* is a cheese-filled mini phyllo pie, sometimes sweetened with honey. We've made it moderately sweet with a cashew–cinnamon–cream cheese–type filling, sprinkled with cinnamon and sugar. You can make them ahead of time and keep them refrigerated (or even frozen!) until ready to bake.

- 1 **package (1 pound, or 454 g) phyllo dough**
- 1 **cup (224 g) nondairy butter, melted**

FOR FILLING:

- 14 **ounces (397 g) silken tofu, drained**
- 1 **cup (112 g) raw cashews**
- 1/4 **cup (84 g) agave nectar**
- 1/4 **teaspoon ground cinnamon**
- 1 **teaspoon pure vanilla extract**
- 2 **tablespoons (30 ml) vegetable oil**
- 8 **leaves fresh mint**
- 1/2 **cup (60 g) walnut pieces**

FOR SUGAR SPRINKLES:

- 2 **tablespoons (24 g) evaporated cane juice**
- 1/8 **teaspoon ground cinnamon**

PREHEAT OVEN to 400°F (200°C, or gas mark 6). Line 2 baking sheets with parchment paper or silicone baking mats.

Thaw phyllo according to package instructions.

Prepare filling by placing tofu, cashews, agave, cinnamon, vanilla, oil, and mint into a blender. Purée until smooth. Stir in walnut pieces.

Unroll phyllo, place long side closest to you. Using a sharp knife or pizza cutter, cut the entire stack in half.

Cut each half in half again, so you end up with strips about 4 inches (10 cm) wide by 12 inches (30 cm) long.

Use about 6 to 8 strips per pie, brushing a layer of melted butter between each strip.

Add about 1/4 cup (60 g) filling to the short end of a butter-brushed pile of strips and fold over one corner to make a triangle. Continue folding pastry strips from side to side in the shape of a triangle until the entire pastry strip covers the filling.

Repeat with remaining pastry strips and filling until all are used.

Place 8 pies on each baking sheet.

Mix together evaporated cane juice and cinnamon for topping.

Brush all pies with an additional melted butter and sprinkle with evaporated cane juice-cinnamon mixture.

Bake for about 30 minutes, or until pastry is puffed, golden, and crispy.

YIELD: 16 pies

PUFF PASTRY FRUIT AND CHEESE DANISHES

By-product sub: nondairy cream cheese for cheese, fruit preserves for honey

The great thing about these delicious Danishes is that you can make as many or as few as you need, refreeze them, and bake them later.

Get creative with the fillings: How about peanut butter and jelly or chocolate?

8   squares (5-inch [13 cm] square) vegan puff pastry,
    thawed according to package instructions
8   tablespoons (1/2 cup, or 120 g) nondairy cream cheese
8   tablespoons (1/2 cup, or 160 g) fruit preserves
    Raw or sparkling sugar for sprinkling

PREHEAT OVEN to 375°F (190°C, or gas mark 5).
Line 2 baking sheets with parchment paper or silicone baking mats.
Lay 1 square of puff pastry on a flat nonstick surface.
Place 1 tablespoon (15 g) cream cheese in the center of the square.
Make an indention in the cream cheese with the back of a spoon.
Place 1 tablespoon (20 g) fruit preserves in the center of the cream cheese.
Fold two opposite corners of the square into the center and pinch together.
Repeat with remaining 2 corners.
Make sure all 4 corners are properly sealed.
Place on baking sheet and repeat with remaining squares.
Sprinkle Danishes liberally with raw or sparkling sugar.
Bake for about 15 to 20 minutes, or until puffed, golden brown, and flaky. The center should pop open revealing the yummy filling.

YIELD: 8 Danishes

MAPLE SPONGE CANDY

 By-product sub: maple syrup for honey

 Corn Free   **For the more experienced cook**   **Gluten Free**

 Nut Free   **Wheat Free**

This is a crispy and sweet, molasses-flavored, optionally chocolate-coated twist on old-fashioned honeycomb candy. Not for the novice, it requires patience and an accurate candy thermometer. Don't let the simple ingredients fool you: This candy is pure science!

> Nonstick cooking spray
> 2 **cups (384 g) Sucanat**
> 1/2 **cup (120 ml) pure maple syrup**
> 1/2 **cup (120 ml) water**
> 1 **tablespoon (14 g) baking soda**
> **Melted chocolate for dipping, optional**

LINE A 9 x 13-inch (23 x 33 cm) baking dish with foil and coat foil with cooking spray.

In a large pot, place the Sucanat, maple syrup, and water. Heat on low, constantly stirring until Sucanat is completely dissolved.

Raise heat and bring to a boil, watching temperature carefully. Bring the temperature to 250°F (121°C) and maintain that temperature constantly for 10 minutes.

Remove from heat and quickly but thoroughly stir in baking soda. The mixture will begin to bubble and froth.

Immediately pour the mixture into prepared dish.

Let cool completely before removing from pan and breaking into pieces.

Dip in melted chocolate, if using. Let cool completely before storing. Store in an airtight container for up to 1 week.

YIELD: 30 pieces

LEMON CAKE

 By-product sub: pectin for gelatin

This old-fashioned cake originally called for a package of lemon Jell-O. Natural fruit pectin, found in the canning section of the supermarket, and lemon-lime soda make a nice stand-in for this sweet and lemony Bundt cake.

You can shake a little powdered sugar on top before serving, or mix up a simple glaze of powdered sugar and lemon-lime soda.

        Nonstick cooking spray
     2 cups (250 g) all-purpose flour
$1^1/2$ teaspoons baking powder
   $^1/2$ teaspoon baking soda
   $^1/4$ teaspoon salt
     1 box (1.75 ounces, or 49 g) fruit pectin, such as Sure-Jel
     1 cup (192 g) evaporated cane juice
     1 6-ounce (170 g) container plain, lemon, or vanilla nondairy yogurt
     1 cup (235 ml) natural lemon-lime soda (not diet!)
   $^1/4$ cup (60 ml) fresh lemon juice
     1 teaspoon pure vanilla extract
     1 tablespoon (15 ml) lemon extract

PREHEAT OVEN to 350°F (180°C, or gas mark 4). Lightly coat a standard bundt pan with cooking spray.

In a large mixing bowl, whisk together flour, baking powder, baking soda, salt, and pectin.

In a medium bowl, beat together evaporated cane juice, yogurt, soda, lemon juice, and extracts.

Add wet to dry ingredients and mix until there are virtually no lumps.

Pour batter into pan, and bake 35 to 45 minutes, or until a toothpick inserted comes out clean.

Let cool before inverting onto a platter.

YIELD: 10 servings

PERFECT CREAM CHEESE FROSTING
WITH ALL-NATURAL FOOD COLORS

 By-product sub: just naturally by-product free!

 Gluten Free   Nut Free   Quick and Easy   Wheat Free

The natural food colors used here yield a pastel look that's not quite as vibrant as commercial food colorings, but they contain no harmful dyes or artificial ingredients.

  1 tub (8 ounces, or 227 g) nondairy cream cheese, such as Tofutti
  1/2 cup (112 g) nondairy butter, cubed
  1 teaspoon vanilla powder
  4 cups (480 g) powdered sugar
     All-natural food coloring, optional

USING AN ELECTRIC MIXER, beat together cream cheese, butter, and vanilla powder until fluffy. Beat in powdered sugar 1/2 cup (60 g) at a time.
   Beat until smooth and fluffy. Beat in desired colors, if using, remembering that using liquid will affect the consistency of the frosting. Add more powdered sugar if necessary.
   Refrigerate until ready to use.

YIELD: Enough for 24 cupcakes or one 9-inch (23 cm) double-layer cake.

Note: Here are the colors you can create using natural ingredients:
- Turmeric: Yellow
- Beet juice or powder: Pink
- Carrot juice or powder: Orange/Peach
- Spinach juice or powder: Green
- Blueberry juice: Purple
- Unsweetened cocoa powder: Chocolate/Brown

SESAME FUDGE

 By-product sub: maple syrup and nondairy milk

 Corn Free   **Wheat Free**

Making candy takes patience. Don't rush the process, although working quickly is a must. We recommend having all of your ingredients ready to go.

Nonstick cooking spray
1 tablespoon (15 ml) pure vanilla extract
2 cups (512 g) thick tahini
1/4 cup (32 g) sesame seeds
2 cups (384 g) evaporated cane juice
1/4 cup (60 ml) pure maple syrup
1/2 cup (120 ml) nondairy milk

LINE A 9-inch (23 cm) square baking dish with foil and coat with cooking spray.

In a medium bowl, mix together vanilla, tahini, and sesame seeds.

In a medium saucepan, heat evaporated cane juice, maple syrup, and milk over low heat until completely dissolved, stirring constantly.

Raise the heat and bring to a full rolling boil, for exactly 1 minute.

Pour boiling mixture into tahini mixture; stir quickly and vigorously to mix.

Immediately pour into prepared baking dish.

Refrigerate to set.

Once completely cooled and hardened, cut into squares.

Store in fridge.

YIELD: 36 pieces

FRUIT JIGGLERS

 By-product sub: agar powder

 Corn Free   **Gluten Free**

 Low Fat   **Nut Free**

 Soy Free   **Wheat Free**

Better than Jell-O! Substitute any juice you have on hand for the açai pomegranate juice. Place whole berries in serving dishes, and pour the mixture on top and surprise your family with real fruit in their dessert.

You can also have fun shaping the dessert; for example, pour mixture into star-shaped ice-cube trays, chill, and have a healthy treat your kids will love!

2 cups (470 ml) açai pomegranate juice blend
1 cup (235 ml) fresh orange juice
2 tablespoons (42 g) agave nectar
1 tablespoon (8 g) agar powder

COMBINE ALL INGREDIENTS in a saucepan. Bring to a boil, whisking constantly; lower heat to medium-low. Cook for another 5 minutes.

Divide into 6 small dessert dishes. Chill for at least 1 hour before enjoying.

YIELD: 6 dessert servings

AMBROS-ISH SALAD

 By-product sub: vegan marshmallows

 Quick and Easy

There are hundreds of variations on this theme, and you can play around with different fruits to add in. For us, it's all about the sweet, creamy, fluffy white dressing. We call for canned fruit in this recipe, but fresh or frozen is certainly okay, too.

12	ounces (340 g) nondairy sour cream, such as Tofutti
1	6-ounce (170 g) container nondairy yogurt
1	tablespoon (15 ml) pure vanilla extract
1	cup (120 g) shredded coconut
3/4	cup (90 g) powdered sugar
1	can (14 ounces, or 397 g) mandarin oranges, drained
1	can (14 ounces, or 397 g) maraschino cherries, drained, stemmed and halved
1	can (14 ounces, or 397 g) pineapple chunks, drained
1	can (14 ounces, or 397 g) peaches, drained
10	ounces (283 g) vegan marshmallows, such as Dandies

WHISK TOGETHER sour cream, yogurt, vanilla, coconut, and powdered sugar.
Fold in fruit and marshmallows.
Refrigerate until ready to serve.

YIELD: 8 to 12 servings

Note: Other favorite fruits to add include grapes, strawberries, blueberries, raspberries, bananas, and mangoes. Some folks like to make this salad even easier by using a big can of fruit cocktail instead of individual cans of fruit.

LET YOUR BODY REJOICE!

FOOLPROOF SUBSTITUTIONS FOR GLUTEN, SOY, SUGAR, AND FAT

Chapter 6

From Crispy Crackers to Rockin' Raw Truffles:
HOW TO SUBSTITUTE FOR GLUTEN

A WORD ABOUT GLUTEN SAFETY: Please be extremely vigilant whether cooking and baking for yourself or another gluten-sensitive person. Check all ingredients thoroughly and be healthy.

Consider the Facts

Gluten is a protein that can be found in all foods that contain wheat, rye, and barley. Ingesting foods that contain gluten can be life-threatening for those who suffer from celiac disease, which prevents vital nutrients from being properly absorbed by the body.

While leading a happy gluten-free life is absolutely possible, it can be tricky because gluten shows up in many processed foods one would assume are safe to consume, such as vinegars, flavor extracts, alcohol, and soy sauce, as well as in cosmetics and medications.

As if it weren't complicated enough already, one also has to take cross-contamination into account: While a gluten-free product should be safe, it won't be if it gets processed on the same equipment as non-gluten-free items. That's why it is so important to be extremely vigilant when purchasing items and ingredients that claim they're free of gluten. Look for brands that certify their product is safe for people who are sensitive to gluten.

Gluten-Free Substitution Guidelines

Cooking gluten-free (GF) foods is easy when it comes to vegetables, fruits, rice, beans, corn, potatoes, and more—all of which are naturally gluten free. Gluten-free baking, on the other hand, can be trickier.

We wish it were as simple as replace "this" with "that," but it isn't. More often than not, it becomes a science experiment in the kitchen, mixing a little bit of this with a little bit of that, to get the perfect crumb, texture, and taste in baked goods. But don't fret; playing the mad scientist every now and again can be fun, and the more you practice, the easier it gets.

For more specific guidelines, let us turn to the following chart, and then apply this knowledge to a nonvegan, nongluten-free recipe.

IF THE ORIGINAL RECIPE CALLS FOR...	REPLACE WITH...
1 cup (125 g) all-purpose or other wheat-based flour	• 1 cup (120 g) Homemade All-Purpose Gluten-Free Baking Mix (page 173) • 1 cup (120 g) store-bought GF baking mix
1 cup (200 g) uncooked pearl barley	• 1 cup (185 g) uncooked brown, white, or wild rice • 1 cup (168 g) uncooked quinoa • 1 cup (200 g) uncooked millet • 1 cup (193 g) uncooked amaranth
1 cup (19 g) puffed wheat	• 1 cup (16 g) puffed brown or regular rice, quinoa, millet, amaranth
8 ounces (227 g) seitan	• 8 ounces (227 g) tofu, tempeh, or TVP

VEGANIZED!: SAMPLE RECIPE

Let's have a look at the following traditional recipe for an example of how we would replace the gluten and nonvegan ingredients:

FUDGY CAPPUCCINO CRINKLES

This fudge-tastic recipe has been adapted from the *Better Homes and Gardens New Cook Book*.

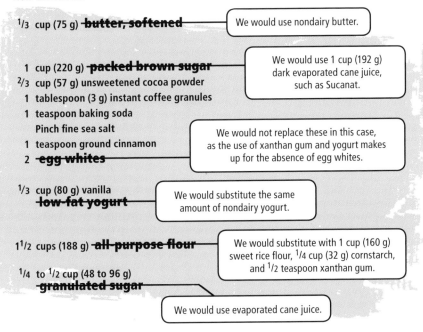

1/3 cup (75 g) ~~butter, softened~~ — We would use nondairy butter.

1 cup (220 g) ~~packed brown sugar~~ — We would use 1 cup (192 g) dark evaporated cane juice, such as Sucanat.

2/3 cup (57 g) unsweetened cocoa powder

1 tablespoon (3 g) instant coffee granules

1 teaspoon baking soda

Pinch fine sea salt

1 teaspoon ground cinnamon

2 ~~egg whites~~ — We would not replace these in this case, as the use of xanthan gum and yogurt makes up for the absence of egg whites.

1/3 cup (80 g) vanilla ~~low-fat yogurt~~ — We would substitute the same amount of nondairy yogurt.

1 1/2 cups (188 g) ~~all-purpose flour~~ — We would substitute with 1 cup (160 g) sweet rice flour, 1/4 cup (32 g) cornstarch, and 1/2 teaspoon xanthan gum.

1/4 to 1/2 cup (48 to 96 g) ~~granulated sugar~~ — We would use evaporated cane juice.

USING AN ELECTRIC MIXER, beat butter with dark evaporated cane juice until creamy. Add cocoa powder, coffee granules, baking soda, salt, and cinnamon. Beat until combined, scraping sides occasionally. Beat in yogurt until combined. Add flour, cornstarch, and xanthan gum. Beat until combined.

Preheat oven to 350°F (180°C, or gas mark 4). Line two cookie sheets with parchment paper or silicone baking mats.

Place evaporated cane juice in a small bowl. Drop 2 heaping teaspoons of dough into sugar and roll into balls. Place 2 inches (5 cm) apart on prepared cookie sheets. Pat down a little. Bake 8 to 10 minutes, or until edges are firm. Transfer to a wire rack as soon as cookies are firm enough to be handled. Let cool completely.

YIELD: 20 cookies

Finding Gluten-Free Substitutes at the Store

Most self-respecting health-food stores are well stocked in gluten-free cooking and baking supplies, and they usually carry many prepackaged crackers, breads, pastas, and snack foods. If you don't have easy access to a health-food store, you should find many items at your regular grocery store or shop online.

Amaranth: Rich in protein and fiber, this nutty-flavored grain cooks up similarly to rice and provides great texture in salads and side dishes. Amaranth flour is great to use in baked goods in combination with other flours and starch.

Brown or white rice flour: Brown rice flour is heartier, has a nuttier flavor, and is richer in nutrients and fiber than white rice flour. Brown makes for crumblier and denser baked goods, while white is better for yielding more delicate results.

Coconut flour: Naturally sweet, makes for rich, high-fiber, high-protein goods. Use in dishes where a hint of coconut flavor is welcome. Works exceptionally well in cookies.

Garbanzo and/or fava bean flour: Excellent source of fiber and protein, has a distinct flavor, almost egg-y.

Gluten-free certified ingredients: Remember to check that commonly used baking ingredients, such as extracts, baking powder, and cornstarch, are certified gluten-free; many are processed with grain alcohol that can contain trace amounts of gluten.

Guar gum: A legume-based, high-fiber thickener that acts as a binder in gluten-free baking. Cheaper than xanthan gum. Use 1 teaspoon per cup (120 g) of flour in bread-type recipes. Use $1/2$ teaspoon per cup (120 g) of flour in muffin-type recipes. Use $1/2$ teaspoon or less or none at all in cookie-type recipes.

Instant ClearJel: Made from modified food starch, usually waxy maize (corn), this pre-gelatinized powder works well to thicken without heat. It should be mixed in with dry ingredients before adding to liquids to prevent lumping.

Quinoa: Great source of calcium, protein, and iron. Nutty flavor. Rinse to get rid of bitter-tasting saponins. Quinoa flour makes for moist and tender baked goods.

Sorghum flour: Millet-like grain that's nutrition-packed. Use in combination with other GF flours in flour mixes to replace wheat flour.

Starches: Arrowroot powder, cornstarch, potato starch, and tapioca starch can all be used with great success in gluten-free baking/cooking. All have a rather neutral flavor, and even though it is said that all have their own specific properties, we find they can be used interchangeably. Note that tapioca flour produces chewier baked goods, making it especially great for use in cookie recipes.

If used in sauces or puddings, it is best to add the starch at the end of cooking, in the form of slurry (mixing equal parts starch with water), and let cook until the mixture thickens. Use 1 tablespoon (8 g) per cup (235 ml) of liquid.

Store-bought gluten-free flour mixes: Bob's Red Mill makes a variety of gluten-free baking mixes, from pancake and waffle mixes to all-purpose baking mixes.

Xanthan Gum: Corn-based thickener that acts as a binder in gluten-free baking. More expensive than guar gum. Use 1 teaspoon per cup (120 g) of flour in bread-type recipes. Use $1/2$ teaspoon per cup (120 g) of flour in muffin-type recipes. Use $1/2$ teaspoon or less or none at all in cookie-type recipes.

Making Gluten-Free Substitutes at Home

Make your own GF flour mixes and you'll get to control which flours go into your mix, as well as the cost.

One thing to bear in mind: The starch (arrowroot, corn, potato, tapioca, etc.) should only represent up to one-third of the total mix.

Using two types of gluten-free flour is best to help with texture and to prevent the flavor of one of flour from overwhelming the other.

It is usually preferable to add xanthan or guar gum when ready to bake rather than adding to the flour mix. It will give you more control over the quantity you need depending on the recipe you've selected.

HOMEMADE ALL-PURPOSE GLUTEN-FREE BAKING MIX

 Gluten Free   Wheat Free

The following is an example of a basic GF baking mix, which is nice to have on hand for quick substitutions in the kitchen. It works well for muffins and cookies and is especially good in deep-frying batters. Use 1 cup (120 g) baking mix to substitute for 1 cup (120 g) wheat flour. Not recommended as a replacement in bread recipes.

$1^{1}/3$ cups (160 g) garbanzo fava bean flour

1 cup (136 g) white sorghum flour

$2/3$ cup (84 g) cornstarch or arrowroot powder

$1/4$ cup (48 g) evaporated cane juice

$3^{1}/2$ teaspoons xanthan gum

$1^{1}/2$ teaspoons salt

$1/2$ teaspoon cream of tartar

COMBINE ALL INGREDIENTS. Store in an airtight container.

YIELD: 4 cups (480 g)

The Recipes: Gluten-Based Recipes ... without the Gluten!

Remember to check that even the most common baking ingredients in your kitchen (baking powder, vinegars, etc.) are free of gluten, and you'll be all set to enjoy and share the delicious fruits of your labor.

SAVORY TRIPLE CORN WAFFLES

 Gluten sub: cornmeal and flaxseed meal

 Gluten Free   **Quick and Easy**   **Wheat Free**

A tasty foundation for dips or salsa. Enjoy with soup or as a snack!

- 4 heaping cups (120 g) gluten-free tortilla chips
- 2 cups (280 g) cornmeal
- 1 cup (104 g) flaxseed meal
- 1 tablespoon plus 1 teaspoon (16 g) baking powder
- 2 teaspoons fine sea salt
- 1 teaspoon cayenne pepper to taste
- 1/4 cup (60 ml) peanut oil
- 1 cup (235 ml) unsweetened nondairy milk, plus extra if needed
- 1 cup (235 ml) green salsa
- 1 cup (164 g) frozen corn
- Nonstick cooking spray

PLACE CHIPS in food processor. Process until mostly ground. Add cornmeal, flaxseed meal, baking powder, salt, and cayenne pepper. Process until combined.

Add oil, milk, and salsa. Process until combined. Add corn, pulse a few times to incorporate. The batter will be very thick, but it shouldn't be too dry; add more milk if needed.

Follow manufacturer's instructions and cook waffles on waffle iron. Spray waffle iron between each waffle for best results.

YIELD: 6 Belgian or 12 standard waffles

STONE-GROUND CORN PANCAKES

 Gluten sub: GF baking mix for flour

 Gluten Free **Nut Free** **Wheat Free**

The corn grits give these pancakes an earthy bite. For an even heartier version, add $1/2$ cup (125 g) corn kernels to the mix. Enjoy with maple syrup or with blueberry compote.

- 1 cup (140 g) dry stone-ground corn grits
- 1 cup (120 g) gluten-free all-purpose baking mix, store-bought or homemade (page 173)
- 2 tablespoons (24 g) evaporated cane juice
- 1 $1/2$ teaspoons baking powder
- $1/2$ teaspoon baking soda
- $1/2$ teaspoon salt
- $1^1/2$ cups (355 ml) nondairy milk
- 2 tablespoons (30 ml) fresh lemon juice
- 2 tablespoons (30 ml) vegetable oil
- 3 tablespoons (30 g) egg replacer powder, such as Ener-G
- $1/4$ cup (60 ml) warm water

IN A MEDIUM mixing bowl, whisk together grits, baking mix, evaporated cane juice, baking powder, baking soda, and salt.

In a small bowl, combine milk and lemon juice; it will curdle and become like buttermilk.

Add oil to the buttermilk mixture.

Whisk together egg replacer powder with warm water until frothy.

Fold into buttermilk and oil mixture.

Fold wet ingredients into dry, being careful not to overmix. The batter should be light and fluffy.

Preheat nonstick pan over high heat.

Prepare as you would any pancake, noting that the cooking time will be a tad longer than traditional pancakes.

YIELD: 8 (6-inch, or 15 cm) pancakes

CHOCOLATE CHIP MUFFINS

 Gluten sub: GF baking mix for flour

 Gluten Free   **Nut Free**

 Wheat Free

These muffins have a great crumb, and you can hardly tell they're gluten free!

2 cups (240 g) Homemade All-Purpose Gluten-Free Baking Mix (page 173)

1/3 cup (27 g) unsweetened cocoa powder

1 teaspoon baking powder

1/2 teaspoon baking soda

1 teaspoon vanilla powder

6 ounces (170 g) nondairy yogurt

1/2 cup (120 ml) canola oil

1 cup (200 g) evaporated cane juice

1/2 cup (120 ml) nondairy milk

1 cup (176 g) semisweet chocolate chips

PREHEAT OVEN to 350°F (180°C, or gas mark 4). Prepare a standard muffin tin with paper liners.

In a large mixing bowl, whisk together Baking Mix, cocoa powder, baking powder, baking soda, and vanilla powder.

In a small bowl, whisk together yogurt, oil, evaporated cane juice, and milk.

Fold wet ingredients into dry, being careful not to overmix.

Fold in chocolate chips.

Fill cupcake liners almost completely.

Bake for 25 to 30 minutes, or until a toothpick inserted in the center comes out clean.

Transfer to wire rack to cool.

YIELD: 12 muffins

RAW FLAX AND HEMP SEED CRACKERS

 Gluten sub: just naturally gluten free!

 Corn Free   **Gluten Free**

 Raw    **Soy Free**

Wheat Free

Make these on your day off, and enjoy the heart-healthy benefits of flax and the essential fatty acids from hemp seed throughout the week.

1/2 cup (84 g) flaxseeds

1/2 cup (80 g) hulled hemp seeds

3 tablespoons (20 g) flaxseed meal

1 tablespoon (6 g) chia seeds

1/4 cup (15 g) chopped fresh parsley

1 tablespoon (15 g) minced garlic

1/2 teaspoon cayenne pepper to taste

1/2 cup (120 ml) water

IN A MEDIUM BOWL, mix together all ingredients and let sit for 1 hour to let the flaxseeds and chia gelatinize.

If you have a dehydrator, spread mixture thinly and evenly on the liner tray.

Alternatively, line a baking sheet with parchment paper and set your oven to the lowest setting, leaving the oven door cracked open to help keep the temperature below 115°F (48°C).

Dehydrate or bake in oven for 4 to 6 hours until crispy. Break into cracker-size pieces.

YIELD: 15 crackers

CURRY TOFU AND ARUGULA SALAD

 Gluten sub: just naturally gluten free!

 Gluten Free **Quick and Easy**

 Wheat Free

The more natural colors on your plate, the healthier the meal! This salad is a riot of peppery green arugula, sweet red peppers, and yellow curry dressing. You're welcome to enjoy it on its own, but why not try scooping it onto a crostini or stuff it in a wrap or pita? If you are not a fan of arugula, feel free to substitute baby spinach.

- 1 pound (454 g) extra-firm tofu, drained and pressed, pan-fried if desired
- 3 ounces (84 g) stemmed arugula
- 1/4 cup (40 g) raisins
- 1/4 cup (30 g) pine nuts
- 1/4 cup (46 g) diced roasted red peppers
  Half a medium red onion, finely diced
- 1 cup (224 g) vegan mayonnaise, store-bought or homemade (page 66 or 67)
- 2 tablespoons (30 ml) apple cider vinegar
- 1 tablespoon (6 g) curry powder to taste
  Salt and pepper to taste

CHOP TOFU INTO small cubes and place in a large mixing bowl.

Add arugula, raisins, pine nuts, red peppers, and onion.

In a small bowl, whisk together mayo, vinegar, curry powder, salt, and pepper.

Add dressing to salad and toss to coat.

YIELD: 4 main-dish or 8 side-dish servings

TEMPEH, CORN, AND ARUGULA SALAD

 Gluten sub: just naturally gluten free!

 Gluten Free **Wheat Free**

This is great on its own as a simple main dish, but it also tastes fantastic in a pita or wrap or piled onto a pizza crust and baked.

- 1 cup (235 ml) vegetable broth
- 8 ounces (227 g) tempeh, cut into bite-size cubes
- 4 cups (2 ounces, or 56 g) baby or wild arugula leaves
- 1/2 cup (84 g) diced pimiento peppers or roasted red peppers
- 1 cup (250 g) corn kernels
- 1 tablespoon (15 ml) sesame oil (not toasted)
- 2 tablespoons (30 ml) tamari
- 1 tablespoon (15 ml) mirin
- 1 teaspoon red pepper flakes
  Salt and pepper to taste

COMBINE BROTH and tempeh in a large skillet. Bring to a boil. Reduce to a simmer; cook until most of the liquid is absorbed.

Add arugula, peppers, and corn to pan. Cook until arugula is wilted, 3 to 5 minutes.

Stir in sesame oil, tamari, mirin, and red pepper flakes.

Cook another 3 minutes, until corn and peppers are heated through.

Add salt and pepper to taste.

YIELD: 2 main-dish or 4 side-dish servings

QUINOA LENTIL PILAF

  Gluten sub: quinoa and lentils

 Corn Free   Gluten Free

 Quick and Easy

 Wheat Free

A mighty tasty combination of gluten-free grains that are great served either cold or warm.

- 1 **cup (170 g) uncooked quinoa, rinsed**
- 1 **cup (192 g) uncooked red lentils, rinsed**
- 2 **medium red bell peppers, cored, seeded, and chopped**
- 1/4 **cup (40 g) raisins or chopped dates**
- 2 **tablespoons (30 ml) extra-virgin olive oil**
- 1/4 **cup (60 ml) fresh orange juice**
- 1/4 **cup (60 ml) apple cider vinegar**
- 1 **teaspoon granulated onion**
- 2 **cloves garlic, peeled and minced**
- 2 **tablespoons (30 ml) tamari**
- 1/2 **teaspoon red pepper flakes to taste**
- 1 **teaspoon caraway seeds**
- 1/2 **teaspoon fine sea salt to taste**
- 1/2 **cup (69 g) chopped, toasted, salted cashews**
- 2 **tablespoons (5 g) chopped fresh basil, optional**

COOK QUINOA and lentils in 2 quarts (1.9 L) water for 10 minutes. Drain and let cool.

Combine remaining ingredients except for cashews and basil in a large bowl.

Add cashews, and sprinkle with basil when ready to serve.

YIELD: 4 servings

YUCCA FRIES

 Gluten sub: just naturally gluten free!

 Corn Free   Gluten Free

 Nut Free   Soy Free

 Wheat Free

Yucca root, or cassava, is a starchy root vegetable that is high in calcium, phosphorous, and vitamin C. You will need to use a knife to peel its thick skin. If you have a deep fryer, this is a great opportunity to use it. Otherwise fill a deep pot with about 4 inches (10 cm) of vegetable oil. Try dipping these fries in Aji Verde (page 208).

- 1 **pound (454 g) yucca root**
- 1/2 **teaspoon ground cumin**
- 1/2 **teaspoon salt**
- 1/2 **teaspoon paprika**
- 1/2 **teaspoon garlic powder**
- 1/2 **teaspoon onion powder**
   Oil for frying

PEEL AND SLICE yucca into fry-size pieces. Rinse under cool water to remove excess starch and prevent discoloring.

Steam fries for about 20 minutes prior to frying to soften and precook them. While yucca is steaming, add spices to a small container with a tight-fitting lid and shake to mix.

Preheat oil to 350°F (180°C). Line a plate or baking sheet with paper towels.

Carefully add steamed fries in small batches to oil. Do not to overcrowd.

Cook for about 3 to 5 minutes, or until golden. Remove from oil, and place on lined tray to absorb excess oil. Sprinkle with seasoning mixture to taste. Serve hot.

YIELD: 4 servings

HORSERADISH, CHIVE, AND POTATO CROQUETTES

 Gluten sub: instant potato flakes for flour

 Corn Free   **Gluten Free**   **Nut Free**   **Wheat Free**

These taste awesome as a side to some "meaty" protein and a leafy green. They also taste great served with sour cream or ketchup.

- 2 tablespoons (30 ml) olive oil
- 1 large leek, whites and light greens only, sliced into thin rings
- 4 cloves garlic, peeled and minced
- 1/4 cup (56 g) nondairy butter
- 1 cup (235 ml) unsweetened nondairy milk
- 1 cup (235 ml) water
- 2^1/2 cups (165 g) instant potato flakes
- 2 tablespoons (30 g) prepared horseradish
- 1 tablespoon (0.5 g) dried chives or 3 tablespoons (3 g) fresh finely chopped chives
- 2 cups (168 g) frozen hash brown potatoes, thawed
  Salt and pepper to taste

PREHEAT OVEN to 425°F (220°C, or gas mark 7).

Line a baking sheet with parchment paper or a silicone baking mat.

Heat oil in a skillet over medium-high heat.

Add leek and garlic; sauté until fragrant and beginning to brown. Remove from heat and set aside.

In a medium pot, bring butter, milk, and water to a boil. Remove from heat and stir in potato flakes.

Add leek and garlic mixture to pot, followed by horseradish, chives, hash brown potatoes, and salt and pepper.

Fill 1/3-cup (80 ml) measuring cup with mixture.

Invert onto baking sheet. Repeat until you have 12 croquettes.

Bake for 15 minutes, flip, and bake for an additional 15 minutes.

Serve warm.

YIELD: 12 croquettes

RED WINE RAGOUT

 Gluten sub: sweet potatoes

 Corn Free Gluten Free Wheat Free

Consider incorporating crispy tempeh or tofu into this tasty gluten-free stew right before serving.

  2 **tablespoons (28 g) nondairy butter**
  2 **tablespoons (30 ml) olive oil**
  4 **medium sweet potatoes, peeled, quartered, and thinly sliced**
  1 **pound (454 g) carrots, thinly sliced into coins**
 1/2 **cup (80 g) chopped onion**
  1 **teaspoon coarse sea salt to taste**
 1/2 **teaspoon ground black pepper**
  2 **teaspoons mild to medium chili powder**
 3/4 **cup (180 ml) red wine**
  2 **tablespoons (30 ml) red wine vinegar**
  2 **tablespoons (30 ml) pure maple syrup**
  3 **dried bay leaves**
  1 **cup (144 g) frozen green peas**

ADD BUTTER and oil to a large pot. Heat over medium heat, until the butter melts. Add potatoes, carrots, onion, salt, and pepper. Cook for 5 minutes. Add chili powder, wine, vinegar, and maple syrup. Stir well.

Bring to a low boil. Burrow bay leaves in the veggies. Cover pot with lid; lower heat to medium-high; and cook for 20 minutes, stirring occasionally.

Add green peas and cook for another 5 minutes, or until veggies are tender.

Discard bay leaves before serving.

YIELD: 4 side-dish servings

INSTANT POTATO CHEESE SOUP

 Gluten sub: just naturally gluten free!

 Gluten Free Quick and Easy Wheat Free

The coolest thing about this soup is how fast it comes together. You can make it even faster by mixing together the dry ingredients ahead of time!

Garnish with a pile of shredded vegan cheese, extra chives, and Imitation Bacon Bits (page 143) for extra excitement. When purchasing instant potato flakes, remember to choose a product with only one ingredient: potatoes.

  5 **cups (1 1/4 quarts, or 1.18 L) vegetable broth**
  2 **tablespoons (36 g) white or yellow miso**

1/4 cup (60 g) nondairy sour cream, store-bought or homemade (page 20)
1 cup (60 g) instant potato flakes
1/4 cup (30 g) nutritional yeast
1 tablespoon (8 g) garlic powder
1 tablespoon (7 g) dried minced onion
1 tablespoon (.5 g) dried chives, or 3 tablespoons (3 g) fresh, finely chopped chives
1/4 teaspoon sweet paprika
Salt and pepper to taste

IN A MEDIUM POT, bring broth to a boil, reduce to a simmer. Add miso, sour cream, potato flakes, nutritional yeast, garlic powder, minced onion, chives, and paprika. Stir until creamy. Season with salt and pepper.

YIELD: 2 large bowls, or 4 side servings

• •

FRENCH ONION SOUP

 Gluten sub: just naturally gluten free!

 Gluten Free   Wheat Free

A classic soup that should be served with a crusty piece of bread (gluten-free bread if preparing a GF-friendly meal) toasted and topped with a slice of melted vegan cheese.

4 whole white or yellow onions, peeled and sliced into thin rings
1/4 cup (60 ml) olive oil
1/4 teaspoon sea salt
1 tablespoon (12 g) evaporated cane juice
2 tablespoons (30 g) minced garlic
2 quarts (1.88 L) vegan beef-flavored broth, such as Better than Bouillon Beef Base

1 dried bay leaf

IN A LARGE POT, preheat oil over medium-low heat.
Add onions and sprinkle with salt.
Caramelize onions by cooking them down for about 35 to 40 minutes, adding the evaporated cane juice about 10 minutes into the process.
Stir often to prevent burning.
The onions should turn translucent, soft, and caramel in color.
Add garlic, broth, and bay leaf.
Bring to a boil; reduce to a simmer.
Cover and simmer for 1 hour.

Note: If you can't get your hands on vegan beef-flavored broth, try this mixture: 1³/4 quarts (1.76 L) low-sodium vegetable broth, ¹/4 cup (60 ml) steak sauce, and ¹/4 cup (60 ml) Bragg Liquid Aminos or gluten-free tamari.

YIELD: 8 cups (2 quarts, or 1.88 L)

PROVENÇAL SOCCA

 Gluten sub: garbanzo bean flour

 Corn Free   **Gluten Free** **Wheat Free**

This traditional crepe-like French delicacy works equally well pizza-style. Yum!

FOR TOPPING:

- 1 **pound (454 g) baby tomatoes, halved**
- 1 **tablespoon (15 ml) olive oil**
   Pinch sea salt
   Pepper to taste
- 1/2 **teaspoon red pepper flakes**
- 1/4 **cup (40 g) chopped red onion**
- 4 **cloves garlic, peeled and minced**
- 12 **basil leaves**
- 4 **sprigs fresh thyme**
- 3/4 **cup (100 g) pitted black olives, halved**
- 2 **sun-dried tomatoes, finely chopped**
- 12 **pitted green olives**
- 1 1/2 **teaspoons capers, without brine**
- 4 **ounces (112 g) crumbled firm tofu, optional**

FOR SOCCAS:

- 2 **tablespoons (30 ml) extra-virgin olive oil**
- 2 **tablespoons (30 ml) apple cider vinegar**
- 3/4 **cup (180 ml) unsweetened nondairy milk**
- 2 **teaspoons Dijon mustard**
- 1/2 **teaspoon sea salt**
   Pepper to taste
- 4 **cloves garlic, peeled and minced**
- 2 **tablespoons (20 g) chopped red onion**
- 8 **torn basil leaves**
- 1 **teaspoon fresh thyme**
- 2 **heaping tablespoons (24 g) dry-roasted almonds, coarsely chopped**
- 1 **cup (120 g) garbanzo bean flour**
- 1/2 **teaspoon baking soda**
- 2 **teaspoons peanut oil**

TO MAKE THE TOPPING: Combine all ingredients in a large saucepan.
   Cook over medium-high heat until tomatoes look deflated, about 8 minutes, stirring occasionally. Set aside.

TO MAKE THE SOCCAS: Combine olive oil, vinegar, and milk in medium bowl. Let stand for 2 minutes then stir in mustard, salt, pepper, garlic, onion, basil, thyme, and almonds. Stir in flour and baking soda until combined. Let stand for 15 minutes.
   Brush a 9-inch (23 cm) skillet with peanut oil. Heat skillet over medium-high heat, add half the batter, spreading until it covers most of the surface. Cook over medium heat for 5 minutes, or until socca is golden brown, carefully flipping it onto a plate and placing it back in the skillet on the other side. Cook for another 5 minutes, until golden brown. Repeat with remaining batter.
   Serve topping alongside soccas, or coat surface of soccas with olive oil, add layer of Roasted Tomato Aioli (page 35) or Triple Olive Spread (page 35), and broil until golden brown.

YIELD: Two soccas

SOUTHWESTERN POLENTA

  Gluten sub: just naturally gluten free!

 Gluten Free   **Nut Free**   **Wheat Free**

This is wonderful served with Nacho Queso (page 52), nondairy sour cream, fresh cilantro, and chopped avocado.

FOR POLENTA:

- 6 **cups (1.4 L) vegetable broth or water**
- 2 **cups (280 g) corn grits**
- 1 **teaspoon ground cumin**
- 1 **tablespoon (15 ml) adobo sauce (from a can of chipotle peppers)**
- 3 **tablespoons (42 g) nondairy butter**

FOR TOPPING:

- 1 **can (15 ounces, or 425 g) black beans, drained and rinsed**
- 1/2 **cup (35 g) diced roasted green chilies**
- 2 **chipotle peppers in adobo, chopped**
- 1 **medium red onion, peeled and diced**
- **One recipe Taco Meat (page 117)**
- 1/2 **cup (4 ounces, or112 g) nondairy cream cheese**

TO MAKE THE POLENTA: Bring the vegetable broth or water to a boil. Reduce heat to low and slowly stir in grits, cumin, and
adobo sauce.

Gently simmer for 30 minutes, stirring often to prevent sticking. It will bubble, so be careful. It will stop bubbling as it thickens, and by the time it is done, it will be very thick with no liquid left. Stir in butter until melted.

Transfer cooked polenta to a 9 x13-inch (23 x 33 cm) baking dish and spread evenly.

Preheat oven to 400°F (200°C, or gas mark 6) while you prepare the topping.

TO MAKE THE TOPPING: In a medium mixing bowl, toss together beans, chilies, peppers, and onion.

Spread evenly on top of polenta.

In a separate bowl, combine Taco Meat and cream cheese; spread evenly on top of bean, chili, and onion mixture.

Bake for 20 minutes, or until toppings are heated through.

YIELD: 10 to 12 servings

SUPERFUDGY COCONUT COOKIE BARS

 Gluten sub: brown rice flour and xanthan gum

 Gluten Free **Wheat Free**

We were told these bars were a hit at a recent Los Angeles vegan bake sale!

- 8 ounces (227 g) finely shredded coconut
- 1 3.5-ounce (100 g) package store-bought roasted chestnuts or pecans
- 1/4 cup (26 g) flaxseed meal
- 1/4 cup (40 g) brown rice flour
- 3/4 cup (144 g) raw sugar
- 1/4 cup (60 ml) peanut oil (reduce to 2 tablespoons [30 ml] if using pecans instead of chestnuts)
- 1/2 cup (120 ml) orange juice
- 2 teaspoons pure vanilla extract
- 1/2 teaspoon pure orange extract
  Nonstick cooking spray
- 1 teaspoon baking powder
- 1/2 teaspoon xanthan gum
- 1/4 teaspoon fine sea salt
- 1/2 cup plus 1 tablespoon (99 g) semisweet chocolate chips

COMBINE COCONUT, chestnuts or pecans, flaxseed meal, flour, and sugar in food processor.

Process until finely ground. Scoop out 1 packed cup (150 g) of mixture.

In a small saucepan, combine 1 cup mixture with oil and juice. Cook over medium-high heat for 2 minutes, until it becomes paste-like. Set aside to cool. Stir in extracts.

Preheat oven to 350°F (180°C, or gas mark 4). Lightly coat an 8-inch (20 cm) square pan with cooking spray.

Add baking powder, xanthan gum, and salt to food processor. Pulse until combined.

Add cooled-down "paste;" process until combined.

Add chips; process until just combined, or transfer to large bowl and combine with a wooden spoon.

Use your hands to press thick mixture into pan.

Bake for 25 minutes, or until edges turn golden brown and batter is set.

Let cool on wire rack for 15 minutes before removing from pan to cool completely. Enjoy chilled.

YIELD: 6 to 8 bars

FLOURLESS PEANUT BUTTER COOKIES

 Gluten sub: flaxseed meal

 Gluten Free **Soy Free** **Wheat Free**

Make your tummy happy by keeping these treats gluten free and flourless.

We haven't tried them with other nut butters, but we're positive they'd be equally delicious.

- 2 cups (512 g) natural crunchy peanut butter
- 1/2 cup (168 g) agave nectar
- 2 teaspoons pure vanilla extract
- 2 teaspoons baking powder
- 1/4 cup (26 g) flaxseed meal
- 1/2 cup (96 g) Sucanat

PREHEAT OVEN to 300°F (150°C, or gas mark 2). Line two large cookie sheets with parchment paper or silicone baking mats.

Combine all ingredients in a small bowl, and stir well for about 1 minute.

Divide dough in 1/4 cup (66 g) portions to get 12 cookies of equal size. Flatten dough to desired shape; cookies won't spread much while baking.

Bake for 20 minutes, or until edges turn golden brown. If cookies are puffy on top, use a spatula and tap gently to flatten.

Let cool on baking sheets until firm enough to transfer to a wire rack to cool completely.

YIELD: 12 large cookies

PEANUT BUTTER CHOCOLATE CHIP COOKIES

 Gluten sub: coconut flour for flour

 Gluten Free **Wheat Free**

We love coconut flour for its soft texture and slightly sweet flavor! The texture of these cookies is reminiscent of a pecan sandie.

    2 cups (512 g) creamy no-stir peanut butter
    1 cup (192 g) evaporated cane juice
    1 cup (112 g) coconut flour
    1 cup (176 g) semisweet chocolate chips
  1/2 cup (120 ml) nondairy milk
    1 tablespoon (15 ml) pure vanilla extract
    1 teaspoon baking powder
    1 teaspoon baking soda
  1/2 teaspoon salt

PREHEAT OVEN to 350°F (180°C, or gas mark 4). Line two cookie sheets with parchment paper or silicone baking mats.

Add all ingredients to a large mixing bowl and combine until well incorporated and crumbly.

Grab about 2 tablespoons (40 g) dough and form into a ball; slightly flatten into a disc and place on cookie sheet.

Place 12 cookies on each sheet and bake for about 15 minutes, or until crackly on top and just beginning to brown.

Let cool at least 5 minutes before transferring to a rack to cool completely.

YIELD: 24 cookies

AMARETTI COOKIES

 Gluten sub: cornstarch or arrowroot powder

 Gluten Free   **Soy Free**   **Wheat Free**

Celine's dad's favorite cookie, made vegan! The texture and flavor are almost like the eggy, crispy, almond-flavored real deal, only sans egg whites—and sans gluten.

- 1¹/₄ cups (200 g) dry-roasted unsalted almonds
- ¹/₂ cup plus 2 tablespoons (120 g) raw sugar
- ²/₃ cup (84 g) cornstarch or arrowroot powder
- 2 teaspoons baking powder
- ¹/₄ teaspoon fine sea salt
- 2 teaspoons pure almond extract
- 2 tablespoons plus 2 teaspoons (40 ml) water

PREHEAT OVEN to 350°F (180°C, or gas mark 4). Line two large cookie sheets with parchment paper or silicone baking mats.

Place almonds, sugar, cornstarch or arrowroot powder, baking powder, and salt in food processor. Process until finely ground.

Add extract and start with 1 tablespoon (15 ml) water, adding more until dough appears a bit dry but holds together when pinched.

Divide dough into 16 equal portions of 1 heaping tablespoon (26 g) each. Do not flatten the cookies as they will spread while baking.

Bake for 18 minutes, or until the edges turn golden brown.

Let cool on sheets until firm enough to transfer to a wire rack. The cookies will crisp up as they cool.

YIELD: 16 cookies

Note: These cookies make an excellent pie crust: Grind 20 of them and combine with 3 tablespoons (45 ml) melted nondairy butter, or enough for the dough to stick together when pinched. Press crust into lightly greased 9-inch (23 cm) round baking pan or pie plate, pour filling on top, and chill before enjoying.

COCOA BARS

 Gluten sub: just naturally gluten free!

 Corn Free   **Gluten Free**   **Soy Free**    **Wheat Free**

A fancy gluten-free treat made of sea salt and chocolate that doesn't require any cooking. These are great for breakfast or as a midday snack, giving you an energy boost whenever you need it.

- 1/2 cup (50 g) pecan halves
- 18 dates (not Medjool), pitted and coarsely chopped
- 1/2 cup plus 2 tablespoons (50 g) unsweetened cocoa powder
  Pinch fine sea salt or fleur de sel
- 3 tablespoons (63 g) agave nectar to taste

GRIND PECANS in food processor until finely ground. Add dates, cocoa, and salt. Process until dates are ground; it might take a few minutes, so patience is key. Add agave and pulse until a paste forms. Sample a little to see if it needs more agave. Press mixture into an 8 x 4-inch (20 x 10 cm) loaf pan lined with parchment paper. Freeze for 1 hour. Slice into desired serving pieces. Store in fridge or freezer.

YIELD: 4 to 6 bars

Variations: This treat offers limitless possibilities! Switch up the type of nut or dried fruit. use carob instead of cocoa powder or don't use either. increase or decrease the amount of dry or liquid sweetener. or for extra crunch. take a tip from our editor and friend Amanda Waddell and add 3/4 cup (14 g) puffed cereal when you're almost done processing the ingredients and moisten the mixture with 2 teaspoons melted coconut oil. You can also use the mixture to shape into balls (as above).

PURELY POUND CAKE

 Gluten sub: xanthan gum, brown rice flour, garbanzo fava bean flour, and egg replacer powder

 Gluten Free **Wheat Free**

A delicious basic cake with a superb texture that invites different flavors to be added by varying extracts, citrus zest, and throwing in chocolate chips or other add-ins. You can also make this recipe in cupcake form; just reduce the baking time by half.

Nonstick cooking spray
3/4 cup (180 ml) nondairy milk
1/2 cup (112 g) nondairy butter, softened
1 cup (192 g) Sucanat
1 tablespoon (15 ml) pure vanilla extract
1/4 teaspoon fine sea salt
1/2 teaspoon xanthan gum
1 cup (160 g) brown rice flour
1/2 cup (60 g) garbanzo fava bean flour
1/2 cup (80 g) Ener-G egg replacer powder
2 teaspoons baking powder

PREHEAT OVEN to 350°F (180°C, or gas mark 4). Lightly coat an 8 x 4-inch (20 x 10 cm) loaf pan with cooking spray.

Blend together milk, butter, Sucanat, vanilla, and salt. The butter might not get thoroughly blended with the rest of the ingredients. Add xanthan gum and blend for 1 minute, until slightly thickened.

In a medium bowl, sift together flours, egg replacer, and baking powder.

Pour wet ingredients into dry and combine until well mixed.

Place batter into prepared pan.

Bake for 40 to 45 minutes, or until a toothpick inserted in the center comes out clean.

Transfer to a wire rack and cool for 15 minutes before removing from pan. Cool completely.

YIELD: One 8-inch (20 cm) loaf

TOFU CRÈME BRÛLÉE

 Gluten sub: just naturally gluten free!

 Corn Free   **Gluten Free**   **Wheat Free**

Experiment with different extracts for exotic flavors. Orange, almond, and chocolate extracts add fabulous flavor—so do espresso powders and dessert liqueurs.

- 8 ounces (227 g) silken tofu, drained but not pressed
- 1/2 cup (120 ml) nondairy creamer
- 1/2 cup (96 g) evaporated cane juice
- 1/4 cup (28 g) raw cashew pieces
- 1 tablespoon (7 g) vanilla powder
- 1/4 cup (48 g) additional evaporated cane juice for topping (vanilla sugar and raw sugar also work wonderfully here)

SOAK CASHEWS for several hours, or overnight. Drain before using.

Preheat oven to 350°F (180°C, or gas mark 4).

Place all ingredients into a blender and purée until very smooth.

Pour evenly into four 3-inch (7.5 cm) ramekins. Place ramekins in the bottom of a baking pan; fill pan with water so that it comes halfway up the sides of the ramekins.

Carefully place pan in oven. Bake for 45 minutes.

Carefully remove pan from oven and refrigerate to cool completely before topping.

Top each ramekin with 1 tablespoon (12 g) sugar, shake gently to evenly distribute.

You can caramelize the sugar by using a kitchen torch, sweeping slowly side to side until browned, or by placing ramekins beneath a preheated broiler and watching closely. Remove from the broiler as soon as tops brown, about 3 minutes.

These are best served immediately after the sugar is caramelized and crispy.

YIELD: 4 servings

SUNFLOWER COCOA BARS WITH MARZIPAN TOPPING

 Gluten sub: sunflower seeds and xanthan gum

 Gluten Free   Wheat Free

These chewy and protein-filled gluten-free bars are great to munch on, come breakfast or snack time.

Nonstick cooking spray
4 cups (567 g) raw sunflower seeds
1 cup (80 g) unsweetened cocoa powder
1¼ cups (240 g) organic brown sugar
1 cup (235 ml) nondairy milk
1 teaspoon baking powder
1 teaspoon xanthan gum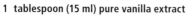
1 tablespoon (15 ml) pure vanilla extract
¼ teaspoon sea salt
¾ cup (180 g) coarsely chopped Homemade Marzipan (page 90)

PREHEAT OVEN to 350°F (180°C, or gas mark 4). Lightly coat a 9-inch (23 cm) square baking pan with cooking spray.

Combine sunflower seeds, cocoa powder, and sugar in a large food processor (or work in several batches if necessary), until seeds are finely ground.

Add milk, baking powder, xanthan gum, vanilla, and salt. Pulse to combine. The batter will be stiff.

Press batter into prepared pan. Sprinkle marzipan over the top, pressing down slightly.

Bake for 25 minutes or until set. Let cool on wire rack. Remove from pan and slice.

YIELD: 12 bars

BANANA UPSIDE-DOWN CAKE

 Gluten sub: GF baking mix for flour

 Gluten Free **Wheat Free**

This cake tastes fantastic served warm with a scoop of vanilla nondairy ice cream.

Nonstick cooking spray

FOR "BOTTOMING":
1 cup (220 g) packed brown sugar
1/4 cup (56 g) nondairy butter
1/2 teaspoon ground cinnamon
2 tablespoons (30 ml) dark rum
2 bananas, ripe but firm

FOR CAKE:
1/2 cup (112 g) nondairy butter
3/4 cup (144 g) evaporated cane juice
1/4 cup (26 g) flaxseed meal
1/3 cup (80 ml) warm water
1 teaspoon pure vanilla extract
1 1/2 cups (180 g) Homemade
   All-Purpose Gluten-Free Baking Mix
   (page 173)
3/4 teaspoon baking powder
1/4 teaspoon baking soda
1/4 teaspoon salt
1/2 cup (120 g) nondairy sour cream
   or yogurt

PREHEAT OVEN to 350°F (180°C, or gas mark 4). Coat a 9-inch (23 cm) cake pan with cooking spray.

TO MAKE THE "BOTTOMING": In a small saucepan, over medium heat, add brown sugar, butter, and cinnamon. Heat until butter is melted and mixture is smooth, stirring often.
   Pour mixture in the bottom of cake pan.
   Slice bananas into 1/4-inch (6 mm) slices and place in a single layer on top of the brown sugar mixture.

TO MAKE THE CAKE: Using an electric mixer, cream together butter and evaporated cane juice, until fluffy.
   Whisk flaxseed meal into warm water and slowly beat into butter and sugar mixture. Add vanilla.
   In a small bowl, combine baking mix, baking powder, baking soda, and salt.
   Add dry mixture to batter in small batches, and mix until well incorporated. Mix in the sour cream.
   Spoon batter on top of bananas and brown sugar mixture, and carefully spread.
   Bake 45 to 55 minutes, or until cake is golden brown and a toothpick inserted in the center comes out clean.
   Remove from oven and cool for 10 minutes.
   Run a sharp knife around the edges of the cake to separate it from the pan. Invert cake onto a serving platter and serve warm.

YIELD: 8 pieces

CRANBERRY ALMOND CUSTARD PIE

 Gluten sub: coconut flour for flour

 Gluten Free **Wheat Free**

This deliciously almond-y concoction is part cheesecake, part fruit crumble, and part pie. It tastes great served with a scoop of nondairy vanilla ice cream.

Nonstick cooking spray

FOR CRUST:

- 1 cup (112 g) coconut flour
- 1/2 cup (112 g) nondairy butter
- 3 tablespoons (36 g) evaporated cane juice

FOR FILLING:

- 8 ounces (227 g) dry-roasted unsalted almonds
- 2 cups (240 g) powdered sugar
- 1/2 cup (112 g) nondairy butter, melted
- 1 tablespoon (15 ml) pure almond extract
- 1 tablespoon (15 ml) pure vanilla extract
- 14 ounces (397 g) silken tofu, drained

- 1 cup (235 ml) nondairy milk
- 1/4 cup (32 g) arrowroot powder or cornstarch
- 1 teaspoon baking soda
- 1 teaspoon baking powder
- 1/4 teaspoon salt
- 1 pound (454 g) fresh or frozen whole cranberries
- 1/4 cup (48 g) evaporated cane juice

PREHEAT OVEN to 350°F (180°C, or gas mark 4). Coat a 9 x 13-inch (23 x 33 cm) baking dish with cooking spray.

TO MAKE THE CRUST: Add coconut flour, butter, and evaporated cane juice to a food processor and process until crumbs form.

Spread crumb mixture along the bottom of the baking dish.

TO MAKE THE FILLING: Add almonds and sugar to a very dry food processor and process until almonds are finely ground. Add melted butter, extracts, tofu, milk, arrowroot or cornstarch, baking soda, baking powder, and salt. Blend until smooth.

Spread mixture evenly over crumb crust. Add cranberries to the top in a single layer. Sprinkle evaporated cane juice evenly over cranberries.

Bake, uncovered, for 40 to 45 minutes, or until filling is bubbling slightly and topping looks crispy.

Place in fridge for 1 hour to set.

YIELD: 12 servings

COCONUT CINNAMON RAISIN BREAD

 Gluten sub: coconut flour for flour

 Soy Free   **Gluten Free**   **Wheat Free**

For coconut lovers only: This superdense and cakelike bread makes use of coconut flour, coconut milk, and shredded coconut. The topping almost tastes like a macaroon! We enjoy it warm at breakfast time, or anytime for a not-too-sweet snack.

FOR BREAD:

1/2 cup (120 ml) warm water

1 1/2 teaspoons active dry yeast

1 tablespoon (12 g) evaporated cane juice

2 cups (224 g) coconut flour

1/2 cup (100 g) evaporated cane juice

1/4 teaspoon salt

1/2 cup (64 g) arrowroot powder or cornstarch

1/4 teaspoon cream of tartar

1/2 teaspoon ground cinnamon

1/2 cup (60 g) finely shredded unsweetened coconut

1/2 cup (120 ml) light coconut milk

1/4 cup (60 ml) vegetable or melted coconut oil

3/4 cup (180 ml) water

1 tablespoon (15 ml) pure vanilla extract

1 cup (160 g) raisins

Nonstick cooking spray

FOR TOPPING:

1/2 cup (120 ml) light coconut milk

1/2 cup (30 g) powdered sugar

1/4 cup (30 g) finely shredded unsweetened coconut

TO MAKE THE BREAD: In a small bowl, add yeast and 1 tablespoon (12 g) evaporated cane juice to the warm water. Mix and set aside to activate.

In a large mixing bowl, combine coconut flour, evaporated cane juice, salt, arrowroot or cornstarch, cream of tartar, cinnamon, and shredded coconut.

Mix coconut milk, oil, water, and vanilla into the yeast mixture.

Add wet ingredients to dry and mix well.

Using your hands, add the raisins and knead everything together.

Cover bowl lightly with a dish towel, and let sit for about 1 hour.

The dough won't rise, but the yeast helps bind and soften it.

Preheat oven to 350°F (180°C, or gas mark 4).

Lightly coat a 9 x 5-inch (23 x 13 cm) loaf pan with cooking spray.

TO MAKE THE TOPPING: Combine coconut milk, powdered sugar, and shredded coconut.

Press dough evenly into prepared loaf pan, making sure to leave at least 1/2 inch (1.27 cm) at the top to account for the topping.

Using a fork, poke holes all over the dough. Pour the topping on top of dough.

Bake for 1 hour, until the topping is slightly browned and a little crackly.

Let cool for at least 15 minutes before slicing.

YIELD: One 9-inch (23 cm) loaf

GF-INGERBREAD CAKE

 Gluten sub: lentils and brown rice flour

 Gluten Free   **Wheat Free**

We're happy to announce that all who have tasted this tender and delicious cake were flabbergasted when they found out it was free of gluten. Imagine their reaction when we announced one of the key ingredients is lentils!

Make it super special by whipping up half a portion of Perfect Cream Cheese Frosting (page 164), adding 1 packed teaspoon lemon zest to it for extra zing.

FOR LENTILS:

1/2 cup (96 g) uncooked red lentils, rinsed

1¹/2 cups (355 ml) water

FOR CAKE:

2 tablespoons (44 g) regular molasses

1/4 cup (56 g) nondairy butter

1/2 cup plus 2 tablespoons (120 g) Sucanat

2 teaspoons ground cinnamon

1 teaspoon ground ginger

1/2 teaspoon ground nutmeg

2 teaspoons pure vanilla extract

1/4 teaspoon fine sea salt

2 tablespoons (10 g) unsweetened cocoa powder

Nonstick cooking spray

3/4 cup plus 2 tablespoons (140 g) brown rice flour

1 teaspoon baking powder

1/4 cup (30 g) powdered sugar

TO MAKE THE LENTILS: Combine lentils and water in a medium saucepan.

Bring to a boil; cook over medium heat about 20 minutes, until water is mostly absorbed and consistency of lentils is purée-like. Drain the lentils.

TO MAKE THE CAKE: Combine molasses, butter, Sucanat, and warm lentils in a medium bowl. Stir until butter melts. Add spices, vanilla, salt, and cocoa powder. Pour mixture into blender, and blend until smooth. Let cool.

Preheat oven to 350°F (180°C, or gas mark 4). Lightly coat an 8-inch (20 cm) square pan with spray.

Add flour and baking powder to mixture and stir until well combined.

Place batter into prepared pan. Bake for 20 to 25 minutes, or until firm and set on top.

Let cool completely on a wire rack before sifting powdered sugar on top to decorate.

YIELD: 6 to 8 servings

Chapter 7
From Cutlets That Definitely Cut It to Savory Waffles:
HOW TO SUBSTITUTE FOR SOY

A WORD ABOUT SAFETY: Please be extremely vigilant whether cooking and baking for yourself or another soy-sensitive person. Check all ingredients thoroughly and be healthy.

Consider the Facts

When following a vegan diet, it is easy to take in more than one's fair share of soy. And while soy has been linked to amazing health benefits, such as having heart-healthy, cholesterol-lowering, and cancer-fighting proteins, overconsumption of the phytoestrogens found in soy have led to claims of early puberty in girls and sexual dysfunction and breast development in boys. Moreover, many people suffer from soy allergies, which explains why all product labels in the United States identify soy as an allergen when it is included as an ingredient.

Moderation is key to a healthy lifestyle, so it is a wise to keep an eye on how much soy one consumes.

Soy Substitution Guidelines

By cutting back on processed and prepackaged foods and focusing on whole, made-from-scratch foods, you will immediately reduce the amount of soy intake in your diet.

One of the easiest ways to cut back or eliminate soy is simply to give up soymilk and tofu. For more specific guidelines, let us turn to the following chart, and then see if we can apply our knowledge to a nonvegan recipe.

IF THE ORIGINAL RECIPE CALLS FOR...	REPLACE WITH...
1 cup (155 g) shelled edamame	• 1 cup (240 g) lima beans, cannellini beans, garbanzo beans, or fava beans
1 cup (235 ml) soymilk	• 1 cup (235 ml) other nondairy milk, such as almond, coconut, rice, oat, or hemp. (Be mindful that no soy is hidden in store-bought nondairy alternatives to soymilk)
8 ounces (227 g) tofu, tempeh, or TVP	• 8 ounces (227 g) seitan, beans, or other legumes (except soybeans or edamame), chickpea tofu, peanut tofu
Soy-based cheese	• Rice-based cheese, Daiya, nutritional yeast-based dishes, or nut-based cheeses, such as Nutty Pepperjack (page 48)

VEGANIZED!: SAMPLE RECIPE

Let's have a look at the following traditional recipe for an example of how we would replace the soy and other nonvegan ingredients:

SESAME STIR-FRY

This recipe has been adapted from the *Better Homes and Gardens New Cook Book*.

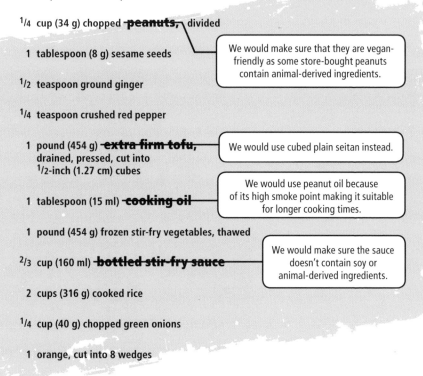

^{1}/4 cup (34 g) chopped ~~peanuts,~~ divided

1 tablespoon (8 g) sesame seeds

> We would make sure that they are vegan-friendly as some store-bought peanuts contain animal-derived ingredients.

^{1}/2 teaspoon ground ginger

^{1}/4 teaspoon crushed red pepper

1 pound (454 g) ~~extra-firm tofu,~~ drained, pressed, cut into ^{1}/2-inch (1.27 cm) cubes

> We would use cubed plain seitan instead.

1 tablespoon (15 ml) ~~cooking oil~~

> We would use peanut oil because of its high smoke point making it suitable for longer cooking times.

1 pound (454 g) frozen stir-fry vegetables, thawed

^{2}/3 cup (160 ml) ~~bottled stir-fry sauce~~

> We would make sure the sauce doesn't contain soy or animal-derived ingredients.

2 cups (316 g) cooked rice

^{1}/4 cup (40 g) chopped green onions

1 orange, cut into 8 wedges

COMBINE 1 TABLESPOON (9 g) peanuts, sesame seeds, ginger, and red pepper in a bowl. Combine with seitan cubes.

Heat oil in a skillet over medium-high heat. Add seitan mixture. Stir-fry for 4 minutes or until mixture browns. Remove from skillet. Add vegetables to skillet, stir-fry until heated through. Add sauce, cook until bubbly. Return seitan to skillet; cook until heated through. Divide rice among 4 serving dishes, spoon seitan preparation over rice. Garnish with remaining peanuts, onions, and orange wedges.

YIELD: 4 servings

Finding Soy Substitutes at the Store

When trying to avoid soy, reading labels is of the utmost importance. Many prepackaged, processed foods contain soy in various forms, so simply shopping for alternative foods, versus looking for direct substitutes for soy, will make your shopping trips much more enjoyable. Shopping the produce department is one of the most obvious places to start. No soy in them apples!

There are a few 1:1 substitutes available, milks being the easiest. Skip the soy and try other milk alternatives: See chapter 1 (page 14) for more information.

A number of manufacturers are recognizing the need for soy-free products. The folks at Earth Balance have released a soy-free version of their buttery spread. Daiya cheeses are both soy-free and gluten-free. South River Miso makes chickpea miso and tamari that is soy-free. The list goes on . . .

Keep an eye out for the many high-protein, soy-free whole foods out there: Seitan, nuts, seeds, legumes, grains, even fruits and vegetables all provide plenty of protein for a soy-free vegan lifestyle.

Making Soy Substitutes at Home

It is unlikely you will be making direct substitutes for soy in your home kitchen since substituting for soy comes down to simply cooking and baking without it. As you become more familiar with your favorite soy-free ingredients, you will undoubtedly be creating wonderful soy-free meals.

See chapter 1 (page 14) for suggested alternatives for soy-based dairy, such as Basic Vanilla Cashew Milk (page 22).

See chapter 2 (page 43) for suggested alternatives for soy-based cheese, such as Nutty Pepperjack (page 48).

See chapter 4 (page 105) for suggested alternatives for soy-based protein sources, such as Pizza Pepperoni Topping (seitan version, page 138).

Recipes without the Soy!

Let's focus on putting these soy facts to good use by whipping up deliciously soy-free creations!

SAVORY SQUASH WAFFLES

 Soy sub: almond milk

 Soy Free

The Mashed Savory Squash makes for scrumptiously spicy waffles that will get even crispier and more delicious after spending a few minutes in the toaster oven:

  1 **cup (244 g) Mashed Savory Squash (page 207), cooled**
1³/4 **cups (415 ml) unsweetened almond milk**
  2 **tablespoons (30 ml) peanut oil**
  2 **teaspoons garam masala**
¹/2 **teaspoon red pepper flakes to taste**
¹/2 **teaspoon fine sea salt to taste**
  2 **cups (240 g) whole wheat pastry flour**
  1 **tablespoon (12 g) baking powder**
    Nonstick cooking spray

COMBINE SQUASH, milk, oil, garam masala, red pepper flakes, and salt in a large bowl. Sift flour and baking powder on top.
    Stir to combine.
    Prepare waffles according to waffle maker manufacturer's instructions.
    Coat waffle iron with cooking spray between each waffle for best results.

YIELD: 5 Belgian or 10 standard waffles

CARROT GINGER MUFFINS

 Soy sub: fruit juice for milk + just naturally soy free!

 Corn Free **Soy Free**

Find yourself longing for autumn? This soy-free recipe makes for perfectly comforting and tender muffins to help tide you over until the leaves start falling again.

    Nonstick cooking spray
1/4 cup plus 2 tablespoons (90 ml) peanut oil
1/2 cup plus 2 tablespoons (150 ml) fresh orange juice
  3 cups (300 g) finely grated carrots
1/2 teaspoon fine sea salt
  1 cup (192 g) Sucanat
  2 teaspoons ground cinnamon
  2 teaspoons pure vanilla extract
  1 tablespoon (8 g) egg replacer powder, such as Ener-G, or 2 tablespoons (12 g) ground flaxseeds
1/4 cup (34 g) crystallized ginger, finely chopped
  2 cups (240 g) whole wheat pastry or all-purpose flour
  2 teaspoons baking powder
  1 teaspoon baking soda

PREHEAT OVEN to 350°F (180°C, or gas mark 4). Lightly coat a jumbo muffin tin with cooking spray.

Simmer oil, juice, carrots, salt, Sucanat, and cinnamon in medium saucepan for 5 minutes. Add mixture to a blender; add vanilla and egg replacer powder or flax; blend. Add chopped ginger. Let the mixture cool completely before using.

Sift flour, baking powder, and baking soda in a large bowl.

Add wet ingredients to dry, being careful not to overmix.

Divide into prepared pan. Bake for 45 to 50 minutes, or until a toothpick inserted into center of muffin comes out clean.

Remove from pan and let cool on a wire rack.

YIELD: 6 jumbo muffins

Note: If you don't have a jumbo muffin tin, bake for 25 minutes in a regular muffin pan.

RAW BANANA APPLE "LEATHER"

 Soy sub: : just naturally soy free!

 **Corn Free** **Gluten Free** **Low Fat**

Nut Free **Raw** **Soy Free** **Wheat Free**

Admittedly, dehydrating can take all day, and it's probably easier just to buy some fruit leather. But if you have 4 different trays in your dehydrator, it's worth preparing a few batches of leathers, Raw Flax and Hemp Seed Crackers (page 176), and other goodies first thing in the morning, set it, check in after 6 to 8 hours and have a week's worth of treats!

2 **large ripe bananas**

1/4 **cup (61 g) apple purée (using applesauce is fine but the recipe would no longer be raw)**

1/4 **teaspoon ground cinnamon**

1 **teaspoon ground vanilla bean**

2 **tablespoons (30 ml) pure maple syrup**

USE A BLENDER or food processor to purée all ingredients.

Spread mixture onto dehydrator's liner tray. If you do not have a liner tray, use parchment paper cut to the shape of your trays.

Set dehydrator to 115°F (48°C) and "cook" for 8 to 10 hours, or until consistency of fruit leather is reached.

If you do not have a dehydrator, use your oven set to the lowest temperature.

Leave the oven door cracked open. Try not to let the oven get above 115°F (48°C).

Spread mixture on a baking sheet lined with parchment paper or a silicone mat.

Using a sharp knife or pizza cutter, cut into 12 strips.

Store between sheets of waxed paper or parchment paper to prevent sticking.

YIELD: 12 strips

SEASONED CROUTONS

 Soy sub: just naturally soy free!

 Soy Free

Flavorful croutons make a tasty addition to any salad, and these crispy bites are a great way to use up stale bread.

    8 slices soy-free sandwich bread
 1/4 cup (60 ml) olive oil
 1/4 cup (21 g) Walnut "Parmesan"
       Sprinkles (page 51)
       Salt and pepper to taste

PREHEAT OVEN to 300°F (150°C, or gas mark 2). Line a baking sheet with parchment paper or a silicone baking mat.

Cut bread into crouton-size cubes and spread in a single layer on the baking sheet. Drizzle oil evenly over bread.

Sprinkle with Parmesan, salt, and pepper.

Bake uncovered for 30 minutes, toss, and bake for an additional 20 minutes.

Let cool completely before storing in an airtight container.

YIELD: 8 servings

MASHED SAVORY SQUASH

 Soy sub: unsweetened almond milk

 Gluten Free   **Soy Free**

 Wheat Free

An alternative to roasting squash that yields an absolutely scrumptious dish that can be used in Savory Squash Waffles (page 204), enjoyed as is, or alongside green vegetables, like Curly Mustard Greens (page 262), and your favorite protein.

    2 tablespoons (30 ml) peanut oil
    1 2 1/2-pound (1.11 kg) butternut
       squash, peeled, quartered,
       and cut into 1/2-inch (1.27 cm) slices
 1/2 cup (80 g) chopped onion
 1/2 teaspoon fine sea salt
    2 large cloves garlic, peeled
and minced
    1 tablespoon (6 g) mild to medium
curry powder
    3 tablespoons (60 g) mango chutney
    1 cup (235 ml) vegetable broth
 1/2 cup (120 ml) unsweetened almond
       milk

HEAT OIL in large saucepan over medium-high heat. Add butternut slices, onion, and salt. Stir often and cook for 10 minutes over medium heat.

Add garlic, curry powder, chutney, broth, and milk. Stir well. Cover with a lid and cook over medium-low heat for 25 minutes, or until squash is tender enough to be mashed. Proceed to mashing.

YIELD: 6 servings

AJI VERDE

 Soy sub: just naturally soy free!

 Corn Free   **Gluten Free**   **Nut Free**   **Quick and Easy**

 Soy Free   **Yeast Free**

Joni's Peruvian friend Sara introduced her to this amazing condiment. It sits in squirt bottles on the tables of Peruvian and South American restaurants and is used for just about everything: to dip sweet potato fries or Yucca Fries (page 178), to pour over bread . . . It even tastes great on salads as a kicky dressing or added to mayo for a tasty aioli.

JONI'S VERSION

2 cups (60 g) fresh baby spinach
Scant 1/2 cup (28 g) fresh parsley
6 aji amarillo escabeche peppers*, stems removed
11/2 cups (355 ml) extra-virgin olive oil
2 tablespoons (30 g) minced garlic
1 tablespoon (15 ml) fresh lemon juice
Salt and pepper to taste

* Aji amarillo escabeche peppers, also known as Peruvian yellow peppers or chile guero, are dark yellowish-orange chili peppers from Peru that are about the size of a large jalapeño. They are most likely found at a Latin market in the freezer section. They are precooked and then frozen.

THE GRINGA VERSION (cheaper, with easier to find peppers!)

2 cups (60 g) fresh baby spinach
Scant 1/2 cup (28 g) fresh parsley
6 roasted jalapeño peppers*
11/2 cups (355 ml) canola oil
2 tablespoons (30 g) minced garlic
1 tablespoon (15 ml) fresh lemon juice
Salt and pepper to taste

*If jalapeños are your pepper of choice, leaving in the seeds or taking them out is up to you. Remove the stems (and seeds if not using) and place on a lined baking sheet and roast at 350°F (180°C, or gas mark 4) for about 30 minutes.

COMBINE ALL INGREDIENTS in a blender and purée until smooth.

YIELD: 2 cups (470 ml)

SPICY HARISSA OVER CROSTINI

  Soy sub: just naturally soy free!

 Corn Free   **Soy Free**

This spicy tomato jam is a nice change from traditional bruschetta. It also tastes wonderful with the Tandoori Spiced Seitan (page 220).

FOR HARISSA:

1 1/2 teaspoons red pepper flakes

1 teaspoon coriander seeds

1 teaspoon cumin seeds

1 teaspoon caraway seeds

2 tablespoons (30 ml) olive oil

2 tablespoons (30 g) minced garlic

2 cans (each 15 ounces, or 425 g) fire-roasted or plain diced tomatoes with juices

1 tablespoon (6 g) fresh chopped mint

Salt and pepper to taste

FOR CROSTINI:

1 French baguette

1/4 cup (60 ml) olive oil

Salt and pepper to taste

TO MAKE THE HARISSA: Grind pepper flakes and coriander, cumin, and caraway seeds in a spice or coffee grinder.

Add oil to the bottom of a medium pot; heat over medium-high heat.

Add garlic and spice mixture; sauté for 1 minute, or until fragrant and spices have absorbed the oil.

Add tomatoes, mint, salt, and pepper.

Reduce heat to low, and simmer, uncovered, for 1 hour or until liquid has evaporated and a thick paste forms. Stir occasionally.

TO MAKE THE CROSTINI: Preheat oven to 350°F (180°C, or gas mark 4).

Slice baguette into 1/4-inch (6 cm)-thick rounds. Arrange in a single layer on a baking sheet and drizzle with olive oil. Sprinkle with salt and pepper.

Bake uncovered, for 20 minutes, or until bread turns golden and crisp.

Serve on a platter with harissa in a dish for dipping.

YIELD: 25 crostini and 3 cups (720 g) harissa for dipping

SMOKY TEQUILA CHILI

 Soy sub: black beans + just naturally soy free!

 Corn Free   Soy Free   Wheat Free

Sit back and enjoy the symphonic results from mixing tequila and orange liqueur with the ancho chili and cumin seeds in this magnificent soy-free chili. It is music to our ears. . . .

FOR ROASTED POTATOES:
- 1 **pound (454 g) baby potatoes, halved**
- 2 **cloves garlic, peeled and quartered**
- 1/2 **teaspoon coarse sea salt**
- 1/2 **teaspoon red pepper flakes**
- 1 **teaspoon cumin seeds**
- 1 **teaspoon ground pepper**
- 2 **tablespoons (30 ml) melted coconut oil**

FOR CHILI:
- 1 **tablespoon (15 ml) peanut oil**
- 4 **cloves garlic, peeled and minced**
- 1 **medium onion, peeled and chopped**
- 4 **ounces (112 g) fire-roasted diced green chilies**
- 1/2 **teaspoon coarse sea salt**
- 3 **tablespoons (23 g) ancho chili powder**
- 1 **tablespoon (6 g) cumin seeds**
- 2 **tablespoons (4 g) dried basil**
- 1/3 **cup (80 ml) tequila**
- 2 **tablespoons (30 ml) triple sec**
- 1 **can (15 ounces, or 425 g) tomato sauce**
- 2 **cans (each 15 ounces, or 425 g) fire-roasted diced tomatoes, with liquid**
- 2 **cans (each 15 ounces, or 425 g) black beans, drained and rinsed**

TO MAKE THE ROASTED POTATOES: Preheat oven to 400°F (200°C, or gas mark 6). Combine all ingredients in a 9-inch (23 cm) baking dish.
Roast for 25 minutes, stirring once halfway through roasting, until tender. Set aside.

TO MAKE THE CHILI: Place oil, garlic, onion, chilies, and salt in large saucepan. Cook over medium heat until onion is tender, about 4 minutes. Stir in chili powder, cumin seeds, and basil. Cook for another minute.
Add tequila and triple sec; bring to a boil. Add tomato sauce and diced tomatoes. Cover and simmer for 15 minutes.
Remove lid, add beans, and simmer uncovered for another 15 minutes, or until chili reaches desired texture. Stir in potatoes, heat through, and serve.

YIELD: 4 to 6 servings

CREAMY EGGPLANT STEW

 Soy sub: chickpeas + just naturally soy free!

 Corn Free   **Gluten Free**

 Soy Free   **Wheat Free**

This creamy and satisfying stew is a delight on cold nights.

- 1 large eggplant, cut into bite-size pieces
- 1/4 cup (40 g) chopped red onion
- 3 cloves garlic, peeled and minced
- 1/2 teaspoon coarse sea salt
- 1/2 teaspoon red pepper flakes
- 4 ounces (112 g) fire-roasted diced green chilies
- 1 can (6 ounces, or 170 g) tomato paste
- 1/4 cup (64 g) crunchy peanut butter
- 1 tablespoon (6 g) garam masala
- 2 1/2 cups (590 ml) water
- 1 can (15 ounces, or 425 g) chickpeas, drained and rinsed, optional
- Handful chopped parsley or scallions

COMBINE EGGPLANT, onion, garlic, and salt in a large saucepan, and cook over medium-high heat until mixture browns and gets fragrant, about 6 minutes. Stir often to prevent burning.

Stir in red pepper flakes, chilies, tomato paste, peanut butter, and garam masala. Cook for another minute. Add water and chickpeas, if using.

Cover, and simmer for 20 minutes, or until eggplant is tender. Garnish with chopped parsley or scallions.

YIELD: 4 main-dish or 8 side-dish servings

TRIPLE-B STEW

 Soy sub: coconut milk and black-eyed peas + just naturally soy free!

 Soy Free

Here is another yummy, exotic, and unusual way to enjoy your protein without having to resort to soy-based meat subs.

- 2 teaspoons peanut oil
- 2 small bananas, cut in 1/2-inch (1.27 cm) chunks
- 2 teaspoons medium chili powder
- 1/4 teaspoon cayenne pepper to taste
- 1/2 teaspoon fine sea salt
- 1/4 cup (40 g) chopped shallots
- 2 large cloves garlic, peeled and minced
- 1/2-inch (1.27 cm) piece fresh ginger, grated
- 1 cup (235 ml) coconut milk
- 1 1/3 cups (312 g) cooked black-eyed peas
- 2 cups (314 g) cooked barley
- 1/2 cup (120 ml) vegetable broth, if needed
- 7 1/2 cups (227 g) fresh baby spinach
- 1/4 cup (4 g) chopped fresh cilantro

ADD OIL to large skillet over medium heat. Add bananas, chili powder, cayenne, salt, shallots, garlic, and ginger and sauté for 2 minutes, until fragrant.

Add coconut milk and simmer over medium heat until milk reduces a bit, about 6 minutes.

Stir in black-eyed peas and barley; simmer for 4 minutes. If preparation thickens up too much, add broth.

Place spinach at the bottom of 4 serving dishes. Divide the hot preparation among all dishes, to wilt spinach. Garnish with cilantro on and serve.

YIELD: 4 servings

TORTILLA SOUP

 Soy sub: just naturally soy free!

 Gluten Free   **Nut Free**   **Soy Free**   **Wheat Free**

Traditional tortilla soup calls for actual tortillas to be cooked into the soup, but if you think about it, what's a tortilla? Masa harina flour mixed with water and a little fat, made into a tortilla and cooked. So why not cut out the middleman and add the masa harina right to the soup? This dish has a sweet tang at first bite, but the heat will sneak up on you. If you are sensitive to spicy foods, fish out the chipotle pepper before the final purée.

Serve garnished with tortilla chips and slices of fresh avocado.

- 2 **tablespoons (30 ml) olive oil**
- 1/2 **white onion, roughly chopped**
- 2 **tablespoons (30 g) minced garlic**
- 1 **can (14 ounces, or 400 g) diced tomatoes with juices**
- 8 **ounces (227 g) tomato sauce**
- 1 **chipotle pepper in adobo sauce**
- 1 **tablespoon (15 ml) adobo sauce from can**
- 2 **teaspoons ground cumin**
- 1 **teaspoon smoked paprika**
- 1/4 **teaspoon ground cinnamon**
- 1/4 **teaspoon unsweetened cocoa powder**
- 2 **cups (470 ml) vegetable broth**
- 1/4 **cup (30 g) masa harina flour**
  Salt and pepper to taste

IN A SOUP POT, heat oil over medium heat.

Add onion and cook for about 5 minutes, until translucent and fragrant.

Add garlic and cook an additional 3 to 4 minutes.

Add tomatoes, tomato sauce, chipotle pepper, adobo sauce, cumin, paprika, cinnamon, cocoa powder, and vegetable broth. Stir and bring to a boil. Reduce to a simmer, cover, and cook for 20 minutes.

Uncover, and stir in masa harina to thicken. Remove from heat. Using a blender, purée mixture until smooth.

YIELD: 4 cups (940 ml) or 2 large servings

GREEN BEANS WITH TOASTED PECAN GRAVY

 Soy sub: pecans and water combination for soy milk + just naturally soy free!

 Soy Free **Wheat Free**

If you like the combination of slightly sweet and salty foods, you'll adore this dish.

1^1/$_4$ **pounds (567 g) fresh green beans, trimmed**
1 **tablespoon (18 g) fine sea salt**
1 **cup (99 g) pecan halves**
1 **tablespoon (15 ml) pure maple syrup**
1 **tablespoon (8 g) cornstarch**
1 **clove garlic, peeled and minced**
1 **teaspoon granulated onion**
^{1}/$_2$ **teaspoon ground white pepper**
^{1}/$_4$ **teaspoon cayenne pepper**
^{1}/$_4$ **teaspoon ground nutmeg**
1 **cup (235 ml) water**

BRING 2 QUARTS (1.88 L) water to a boil. Add green beans, 2 teaspoons of salt, and cook over medium-high heat for 6 minutes, or until beans are crisp tender.

Drain, then plunge beans into ice-cold water.

Preheat oven to 350°F (180°C, or gas mark 4). Combine pecans, syrup, and 1 teaspoon salt in a small baking dish. Toast in oven for 10 minutes.

In medium bowl, combine cornstarch, garlic, onion, white pepper, cayenne pepper, and nutmeg. Stir in water. Add toasted pecans. Using an immersion blender, blend until smooth.

Cook over medium heat in a medium saucepan, until thickened, about 4 minutes.

Serve warm gravy on top of beans.

YIELD: 4 side-dish servings; 1^1/$_2$ cups (355 ml) gravy

LINGUINE WITH GREEN BEANS AND SPINACH WALNUT PESTO

 Soy sub: just naturally soy free!

 Soy Free

This is fresh and light and unlike many pasta dishes. Enjoy as a main dish, a side dish, or replace the linguine with spirals and serve as a cold pasta salad.

1 **pound (454 g) linguine, prepared according to package instructions**
3 **cups (340 g) fresh green beans, trimmed and steamed to desired tenderness**

FOR PESTO:
^{1}/$_2$ **cup (120 ml) olive oil**
1 **medium white onion, coarsely chopped**
2 **tablespoons (30 g) minced garlic**
2 **cups (60 g) fresh baby spinach**
^{1}/$_2$ **cup (30 g) fresh curly parsley, stems removed**
1 **cup (120 g) walnut pieces**
2 **tablespoons (30 ml) fresh lemon juice**

WHILE PASTA is boiling and green beans are steaming, prepare the pesto.

Add 2 tablespoons (30 ml) of olive oil to a medium skillet and heat over medium heat.

Add onion and garlic and sauté until fragrant and translucent.

To a food processor, add sautéed onion and garlic, spinach, parsley, walnuts, lemon juice, and remaining olive oil.

Purée until well chopped but still a little chunky.

After pasta is cooked, drain and return to the pot. Toss with pesto and green beans.

YIELD: 4 main-dish or 8 side-dish servings

RUSTIC PASTA BAKE

 Soy sub: just naturally soy free!

 Soy Free

Chock-full of vegetables, this simple pasta dish boasts some serious flavor. Slightly Mediterranean, it tastes great hot but will win you over if you try it cold the next day.

Make it gluten free by replacing the pasta with quinoa pasta and using GF balsamic vinegar.

> 1 pound (454 g) pasta, cooked according to package directions
> 1 can (15 ounces, or 425 g) white beans, drained and rinsed
> 1 can (15 ounces, or 425 g) diced tomatoes with juices
> 1 can (15 ounces, or 425 g) artichoke hearts, quartered
> 2 cups (142 g) chopped broccoli florets
> 1 cup (160 g) diced white onion
> 1/2 cup (60 g) pine nuts
> 1/4 cup (60 ml) olive oil
> 3 tablespoons (45 ml) balsamic vinegar
> 3 tablespoons (45 g) julienne-cut, oil-packed sun-dried tomatoes
> 2 tablespoons (30 g) minced garlic
> 2 tablespoons (30 g) mild Dijon mustard
> 1 tablespoon (2 g) dried basil or 3 tablespoons (9 g) fresh chopped basil
>   Salt and pepper to taste

PREHEAT OVEN to 375°F (190°C, or gas mark 5).

In a large mixing bowl, combine all ingredients. Stir in cooked pasta.

Spread mixture in 9 x 13-inch (23 x 33 cm) casserole dish. Bake, covered, for 45 minutes, or until top is slightly browned and crisp.

YIELD: 8 servings

BEET RICE SALAD

 Soy sub: black-eyed peas + just naturally soy free!

 Corn Free   **Soy Free**   **Wheat Free**

This salad is not only a complete protein thanks to the combination of beans and rice, but its glorious scarlet color comes compliments of the beets!

For a brighter pink hue, skip the orange juice, blend one 15-ounce (425 g) can beets including ¼ cup (60 ml) of the liquid, and add 2 packed teaspoons orange zest to the salad.

  2 cups (470 ml) water
  1 cup (185 g) uncooked brown jasmine rice
  1 cup (170 g) cooked, sliced baby beets
¼ cup (60 ml) red wine vinegar
¼ cup (60 ml) fresh orange juice
¼ cup (60 ml) extra-virgin olive oil
  2 teaspoons mild Dijon mustard
  1 teaspoon fine sea salt to taste
½ teaspoon ground white pepper to taste
  1 large clove garlic, peeled and minced
  2 teaspoons ground coriander
¼ cup (40 g) chopped red onion
1⅓ cups (312 g) cooked black-eyed peas, drained and rinsed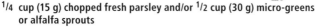
¼ cup (15 g) chopped fresh parsley and/or ½ cup (30 g) micro-greens
     or alfalfa sprouts

BRING WATER to a boil in medium saucepan. Stir in rice, cover with a lid.

Lower heat to medium. Cook rice until water is absorbed and rice is tender, about 25 minutes. Chill rice in fridge.

Using a blender, blend together beets, vinegar, orange juice, oil, mustard, salt, pepper, garlic, and coriander until smooth.

Combine chilled rice, beet dressing, onion, black-eyed peas, parsley, and/or micro-greens or sprouts. Enjoy at room temperature or chilled.

YIELD: 4 servings

BARLEY, SPINACH, AND WALNUT SALAD

 Soy sub: just naturally soy free!

 Corn Free   Low Fat

 Soy Free   Wheat Free

Feel free to halve this recipe unless you plan on taking it to a potluck. And, it's great to make this a day or two ahead to allow the spinach to wilt and the flavors to meld.

- 1 cup (200 g) uncooked pearl barley
- 3 cups (705 ml) lightly salted water
- 1/4 cup (60 ml) balsamic vinegar
- 2 tablespoons (30 g) minced garlic
- 2 tablespoons (30 g) mild Dijon mustard
- 2 tablespoons (30 ml) fresh lemon juice
- 1 tablespoon (21 g) agave nectar
- 1 shallot, finely diced
- 3 cups (90 g) baby spinach
- 8 ounces (227 g) fresh mushrooms, sliced or chopped
- 1/2 cup (60 g) walnut pieces
  Salt and pepper to taste

There are a few ways to prepare the barley. Mix together water and barley in lightly salted water in an oven-safe casserole, cover and bake at 375°F (190°C, or gas mark 5) for 45 minutes, or until liquid is absorbed. You can also use a rice cooker, or a pot on the stovetop. When barley is cooked, set aside to cool.

For dressing, whisk together vinegar, garlic, mustard, lemon juice, agave, and shallot. Set aside.

In a large mixing bowl, toss together cooled barley, spinach, mushrooms, and walnuts. Add dressing and toss to coat.

YIELD: 8 to 12 servings

LIMAS FOR LIMA LOVERS

 Soy sub: just naturally soy free!

 Corn Free   Gluten Free

 Nut Free   Quick and Easy

 Soy Free   Wheat Free

This simple preparation showcases the lima for the lima's sake. If possible, use fresh herbs (triple the amount of dried) for a brighter flavor.

- 1 pound (454 g) fresh or frozen lima beans
- 2 tablespoons (30 ml) olive oil
- 1 tablespoon (15 g) minced garlic
- 1/4 cup (40 g) diced yellow or white onion
- 1 teaspoon dried thyme
- 1 teaspoon dried oregano
- 1 teaspoon dried parsley

Steam or boil beans to desired tenderness in lightly salted water.

While beans are cooking, heat oil over medium-high heat and sauté garlic and onion until fragrant and translucent, 3 to 5 minutes.

Once beans are ready, drain and add to garlic and onion mixture.

Stir in herbs.

YIELD: 4 servings

TOMATO CAPER COUSCOUS

 Soy sub: just naturally soy free!

 Corn Free   Nut Free

 Quick and Easy   Soy Free

Thought you didn't like capers? Let this couscous prove you wrong. And you may find yourself wanting to use this soy-free dressing with other grains, too!

FOR DRESSING:

- 2 tablespoons (15 g) capers with brine
- 1 tablespoon (15 g) whole-grain Dijon mustard
- 1/4 cup (60 ml) dry white wine
  Pepper to taste
- 1/2 teaspoon fine sea salt
- 2 tablespoons (30 ml) extra-virgin olive oil
- 3 medium Roma tomatoes, chopped
- 2 cloves garlic, peeled and minced
- 1 teaspoon granulated onion

FOR COUSCOUS:

- 1 cup (235 ml) water
- 1 cup (186 g) whole wheat couscous
- 1/2 teaspoon fine sea salt
- 2 generous handfuls fresh baby spinach

TO MAKE THE DRESSING: Combine capers, mustard, wine, pepper, salt, oil, tomatoes, garlic, and onion in a medium saucepan. Bring to a boil, lower heat.
Simmer for 10 minutes. Crush tomatoes with a potato masher; set aside.

TO MAKE THE COUSCOUS: Bring water to a boil, remove from heat, stir in couscous, cover and let steam for 5 minutes. Fluff with a fork. Let cool. Combine couscous with dressing and baby spinach. Serve at room temperature.

YIELD: 4 side-dish servings

RED THAI COCONUT CURRY RICE

 Soy sub: just naturally soy free!

 Corn Free   Gluten Free

 Quick and Easy   Soy Free

 Wheat Free

This rich and creamy rice tastes great on its own or alongside your favorite protein as a side dish.

- 1 can (14 ounces, or 414 ml) full-fat coconut milk
- 2 cups (470 ml) water
- 3 tablespoons (45 g) Thai-style red curry paste
- 2 tablespoons (32 g) creamy no-stir peanut butter
- 2 tablespoons (30 g) minced garlic
- 1 cup (180 g) uncooked jasmine rice
- 2 tablespoons (2 g) fresh chives, finely chopped
  Salt and pepper to taste

In a medium pot with a tight-fitting lid, bring coconut milk, water, and curry paste to a boil, reduce to a simmer.
Stir in peanut butter and garlic.
Stir in rice, cover, and cook for about 10 minutes or until rice is tender, stirring often.
Remove from heat and let sit for about 10 minutes to absorb excess liquid.
Stir in chives, salt, and pepper.

YIELD: 4 servings

TANDOORI SPICED SEITAN

 Soy sub: just naturally soy free!

 Corn Free   **Soy Free**

This tasty seitan is a natural to pair with your favorite Middle Eastern dishes. It's also great on its own or on a sandwich with a handful of peppery arugula!

1 cup (144 g) vital wheat gluten flour
$^1/_2$ cup (60 g) whole wheat pastry flour
1 tablespoon (8 g) garlic powder
1 tablespoon (14 g) tightly packed brown sugar
$^1/_2$ teaspoon ground coriander
$^1/_2$ teaspoon ground cumin
$^1/_4$ teaspoon ground cardamom
$^1/_4$ teaspoon ground cinnamon
$^1/_4$ teaspoon salt
$^1/_8$ teaspoon ground black pepper
$1^1/_2$ cups (353 ml) water
2 tablespoons (30 ml) olive oil
Aluminum foil

PREHEAT OVEN to 350°F (180°C, or gas mark 4).

In a mixing bowl, whisk together flours and spices. Add water and oil and mix until well incorporated.

Transfer dough into the center of a large sheet of aluminum foil. Roll up the foil, and twist the ends tight to seal.

Bake, seam side down, on a baking sheet or in a baking dish for 90 minutes, or until firm.

YIELD: 8 servings

WHITE BEAN CUTLETS

 Soy sub: just naturally soy free!

 Corn Free   **Soy Free**

These easy-to-make flaky cutlets are reminiscent of fish fillets. Eating gluten free? Just replace the panko with GF bread crumbs.

1 can (15 ounces, or 425 g) white beans, drained and rinsed
1 tablespoon (15 g) minced garlic
1 tablespoon (2 g) dried parsley
$^1/_2$ teaspoon paprika
$^1/_2$ teaspoon celery salt
$^1/_4$ teaspoon ground black pepper
$^1/_4$ cup (60 ml) vegetable oil
1 cup (80 g) panko bread crumbs
$^1/_4$ cup (60 ml) water, as needed
2 tablespoons (30 ml) additional oil for frying

ADD BEANS, garlic, parsley, paprika, celery salt, pepper, and oil to a large mixing bowl and mash together until there are almost no whole beans remaining. You can use a potato masher but we find using our hands works really well.

Add panko and mix well. Let stand for about 20 minutes.

Divide mixture into 4 equal portions, and form into cutlets about 3 inches (7.5 cm) wide, 4 inches (10 cm) long and $^1/_3$ inch (1 cm) thick. If your mixture is too dry and crumbly, add more water, a little at a time.

Preheat oil in a skillet over medium heat. Pan fry cutlets for 3 to 5 minutes per side, or until golden and crispy.

YIELD: 4 cutlets

MUSHROOM MARGARITA FAJITAS

 Soy sub: mushrooms for tofu or other soy-based proteins + just naturally soy free!

 Corn Free   **Gluten Free**    **Low Fat**

Soy Free   **Wheat Free**

These citrusy-spicy fajitas make for a light yet filling meal. Serve garnished with fresh cilantro and slices of ripe avocado for a real fresh flavor.

FOR MARINADE:

- 1/4 **cup (60 ml) fresh lemon juice**
- 1/4 **cup (60 ml) fresh lime juice**
- 2 **tablespoons (30 ml) tequila**
- 1 **tablespoon (12 g) evaporated cane juice**
- 1/2 **teaspoon garlic powder**
- 1/2 **teaspoon onion powder**
- 1/2 **teaspoon chili powder**
- 1/2 **teaspoon sweet or smoked paprika**
- 1/4 **teaspoon cayenne pepper**
- 1/4 **teaspoon sea salt**
- 1/4 **teaspoon ground black pepper**

FOR FAJITAS:

- 2 **large portobello mushrooms, stems removed**
- 1 **red bell pepper**
- 1 **green bell pepper**
- 1 **medium red or white onion**
- 8 **fajita-size tortillas**

TO MAKE THE MARINADE: Whisk all ingredients together and set aside.

TO MAKE THE FAJITAS: Cut portobellos into thin strips.
   Core and seed peppers, and cut into thin strips.
   Roughly chop the onion.
   Place vegetables in a large resealable plastic bag, or a container with a tight-fitting lid.
   Pour marinade over vegetables and shake to coat.
   Marinate for a few hours, or overnight.
   To cook, preheat a large skillet or grill pan over high heat and sauté vegetables until they reach desired tenderness.
   Serve sizzling hot with warm tortillas.

YIELD: 8 fajitas

POTATO CHUTNEY OVER RICE

 Soy sub: just naturally soy free!

 Soy Free

This warm and comforting dish makes a great one-bowl meal, or round it out by serving with a hearty side of steamed kale and chickpeas.

 2   medium russet potatoes with skins, cubed (4 cups [470 g])
 3   cups (705 ml) vegetable broth
 1   shallot, diced
 3   cloves garlic, peeled and chopped
 2   tablespoons (30 ml) olive oil or nondairy butter
 5   dried apricots, chopped
 $^1/_3$ cup (54 g) raisins, chopped
 $^1/_3$ cup (40 g) chopped walnuts
 $^1/_2$ teaspoon ground cumin
 $^1/_4$ teaspoon ground allspice
 $^1/_4$ teaspoon ground cinnamon
 $^1/_4$ teaspoon garam masala
 $^1/_4$ teaspoon curry powder
 2   tablespoons (42 g) agave nectar
 2   tablespoons (30 ml) apple cider vinegar
     Salt and pepper to taste
 4   cups (474 g) cooked rice

RINSE POTATOES under cool water to remove excess starch and prevent discoloring.

To a large skillet, add vegetable broth. Bring to a boil.

In a separate pan, over medium-high heat, sauté shallot and garlic in oil or melted butter for 2 to 3 minutes, until fragrant and slightly browned.

Add potatoes to boiling broth. Return to boil and reduce to simmer.

Add sautéed mixture, along with any excess oil or butter, to the potatoes.

Stir in dried fruit, nuts, and spices.

Simmer uncovered until almost all liquid is absorbed, about 15 minutes.

Stir in agave and vinegar.

Add salt and pepper to taste.

Spoon mixture over rice and serve.

YIELD: 4 servings

Chapter 8

From Naturally Sweet Pies to Crazy Caramel Baked Apples:
HOW TO SUBSTITUTE FOR REFINED SUGAR

THE PROMISE OF SOMETHING SWEET to seal the end of an already tasty meal never sounded as good as it does nowadays with the many all natural sweeteners available at the market, many of which you probably already stock in your pantry. Let's start with a little bit of history...

Consider the Facts

Blame Columbus if you have a ferocious sweet tooth, because he's the one who brought sugar to the Americas. However, back in his time, sugar was hardly as refined as it is today. The process was kept rather simple: Boil the cane juice until evaporation occurs, leaving deliciously sweet crystals behind that retained some of the essential nutrients. Nowadays? Not so simple.

The process of refining white sugar strips the naturally occurring phosphorus, calcium, iron, magnesium, and potassium found in the sugar cane plant, turning a naturally sweet treat into a bone-char bleached mess of empty calories. Research shows that refined white sugar is linked to tooth decay, weight gain, diabetes, and many other health problems.

What's more, many food companies are replacing plain white sugar with chemically-altered high-fructose corn syrup because it is cheaper. Aside from having no taste, the chemical process involved in making high-fructose corn syrup strips out any nutritional value whatsoever. Its altered chemical structure prevents our bodies from breaking it down, and that can wreak havoc on the pancreas.

Enter artificial sweeteners, but at what cost? Saccharin, aspartame, acesulfame-K, neotame, and most recently, sucralose, all have had their 15 minutes of fame, promising to slim our waistlines by providing sweetness without adding calories. The FDA would have us believe these non-nutritive sweeteners are safe. While scientific studies have concluded that the link between cancer and artificial sweeteners is unfounded, we prefer to maintain a diet free of anything artificial while focusing on using healthier sweeteners in moderation.

Luckily, there are many options available for those who choose not to add refined sugar to their diet, which we will describe in the "Finding Refined Sugar Substitutes at the Store" section on page 226.

IF THE ORIGINAL RECIPE CALLS FOR...	REPLACE WITH...
1 cup (225 g) brown sugar, and you want to switch to a similar but healthier dry sweetener or liquid sweeteners	• 1 cup (192 g) dark evaporated cane juice, such as Sucanat • 1 cup (235 ml) liquid sweetener (e.g., agave nectar, barley malt, pure maple syrup, blackstrap or regular molasses, brown rice syrup, fruit syrup), decrease other liquid by 1/3 cup (80 ml) or add extra 1/4 cup (30 g) flour to baked good recipe
1 cup (192 g) white sugar, and you want to switch to a similar but healthier dry sweetener or liquid sweeteners	• 1 cup (192 g) evaporated cane juice, such as Florida Crystals • 1 cup (235 ml) liquid sweetener (e.g., agave nectar, barley malt, pure maple syrup, blackstrap or regular molasses, brown rice syrup, fruit syrup), decrease other liquid by 1/3 cup (80 ml) or add extra 1/4 cup (30 g) flour to baked good recipe
1 cup (235 ml) liquid sweetener (e.g., high-fructose corn syrup), and you want to switch to a similar but healthier liquid sweetener or dry sweeteners	• 1 cup (235 ml) agave nectar, barley malt, pure maple syrup, blackstrap or regular molasses, brown rice syrup, fruit syrup • Fruit purée or fruit juice concentrate to sweeten nondairy yogurts and smoothies, quantity to taste • 1 cup (192 g) dry sweetener, increase liquid by 1/3 cup (80 ml) or remove 1/4 cup (30 g) flour from baked good recipe
1 cup (120 g) powdered sugar	• 1/2 cup plus 2 tablespoons (120 g) evaporated cane juice ground in a very dry blender until powdery • 1 cup (120 g) store-bought organic powdered sugar

VEGANIZED!: SAMPLE RECIPE

Let's have a look at the following traditional recipe for an example of how we would replace the refined sugar and other nonvegan ingredients:

POPCORN AND CANDY BALLS

This treat has been adapted from the *Better Homes and Gardens New Cook Book*.

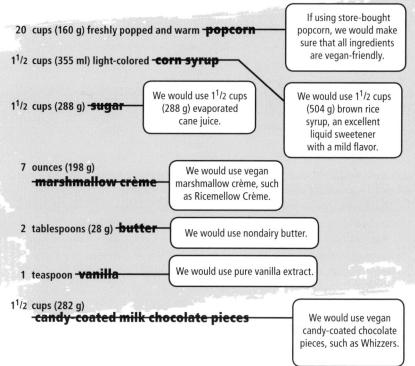

20 cups (160 g) freshly popped and warm ~~popcorn~~

> If using store-bought popcorn, we would make sure that all ingredients are vegan-friendly.

1¹/₂ cups (355 ml) light-colored ~~corn syrup~~

> We would use 1¹/₂ cups (504 g) brown rice syrup, an excellent liquid sweetener with a mild flavor.

1¹/₂ cups (288 g) ~~sugar~~

> We would use 1¹/₂ cups (288 g) evaporated cane juice.

7 ounces (198 g) ~~marshmallow crème~~

> We would use vegan marshmallow crème, such as Ricemellow Crème.

2 tablespoons (28 g) ~~butter~~

> We would use nondairy butter.

1 teaspoon ~~vanilla~~

> We would use pure vanilla extract.

1¹/₂ cups (282 g) ~~candy-coated milk chocolate pieces~~

> We would use vegan candy-coated chocolate pieces, such as Whizzers.

PLACE WARM POPCORN in a very large heatproof bowl, set aside.

In a medium saucepan, bring syrup and evaporated cane juice to a boil over medium-high heat, stirring constantly.

Remove from heat. Combine with marshmallow crème, butter, and vanilla.

Pour marshmallow mixture over popcorn, stirring to coat. Cool until popcorn can be handled. Stir in chocolate pieces. Using slightly dampened hands, shape mixture into 3-inch (8 cm) balls. Wrap individually in plastic. Store at room temperature for up to a week.

YIELD: 16 popcorn balls

Finding Refined Sugar Substitutes at the Store

Almost every grocery store should carry at least one type of healthy sweetener to help you eliminate refined sugars from your diet, be it in the baking aisle, international foods aisle, or breakfast foods aisle. If you cannot find what you're looking for, shopping online is another option.

Agave nectar or **syrup:** Made from the juice of agave, a succulent plant. Perfect as a substitute for honey. Reduce oven temperature by 25°F (3.9°C) when baking treats that include this sweetener, as it tends to over-brown easily.

Barley malt syrup: Produced from sprouted barley, this syrup is thick and dark with a distinct malty, almost molasses-like flavor. Because it is a complex sugar, it breaks down slowly in the body, preventing that sugar "high" caused by refined sugars.

Blackstrap molasses: This dark and bittersweet syrup is what is left over once cane juice has been crystallized. It is rich in calcium and iron. Regular molasses can be used in its place, but it provides less nutritional benefits.

Brown rice syrup: This all-natural liquid sweetener has the consistency of honey. It is made by combining barley malt and brown rice and cooking the mixture until the starches are converted to sugar. It is then strained and reduced to a thick syrup. It has a unique caramel-like flavor that can enhance baking recipes without making them too sweet.

Date sugar: Natural sweetener obtained from dates. We use it in cooking applications rather than for baking; it's also good for adding to breakfast cereal and beverages.

Evaporated cane juice: Tastes, measures, and performs exactly like refined white sugar in recipes. Its crystals can be a bit coarser, depending on the brand. Brands like Wholesome Sweeteners (which is also fair trade certified), Florida Crystal, Rapadura, and Zulka are all fairly easy to find.

Palm sugar: Natural sap collected from sugar palms, with a flavor reminiscent of molasses. We use it in cooking applications rather than for baking and also add it to breakfast cereal and beverages.

Pure maple syrup: The sweet sap adds a boost of manganese and zinc to your morning pancakes but also works well in many other cooking and baking applications. We prefer the grade B dark amber variety for its deeper flavor and higher concentration of nutrients.

Sugar cane natural, or Sucanat: A darker-colored type of evaporated cane juice with a strong molasses flavor. It delivers iron, calcium, potassium, and other minerals.

Low-Calorie All-Natural Sweeteners:

Pure stevia: An all-natural, calorie-free sweetener that comes from the tropical stevia plant native to South America. Approximately 300 times sweeter than sugar, so a little goes a long way.

Pure xylitol: Xylitol is a naturally occurring sweetener found in the fibers of many fruits and vegetables, including various berries, cornhusks, oats, and mushrooms. It measures 1:1 for sugar. Xylitol is not calorie-free; however, it contains 40 percent fewer calories than sugar and is absorbed more slowly by the digestive system, limiting its impact on blood-sugar levels. One thing to remember when cooking and baking with xylitol is that is highly absorbent, so you may need to adjust the amount of liquid you use.

Z-Sweet and **Truvia** are blends of both stevia and xylitol, with additional filler to help bulk them up and make them easier to use in recipes.

As is the case with all foods and especially those that are fairly new to the market, we recommend using moderation when consuming these products. Have a look at the following chart to help you replace sugar with these sweeteners in your recipes, as they usually do not measure 1:1 for sugar.

Refined Sugar Substitutes

Refined White Sugar	Stevia/Xylitol Blends (Packets)	Stevia/ Xylitol Blends (Bulk)	Clear Stevia Liquid	Pure Stevia Powder	Pure Xylitol
2 teaspoons	1 packet	1/2 teaspoon	1/4 teaspoon	1/16 teaspoon	2 teaspoons
1 cup (192 g)	24 packets	1/4 cup (48 g)	2 1/2 teaspoons	1 1/2 teaspoons	1 cup (192 g)

Making Sugar Alternatives and Substitutes at Home

Here are a few tips on how to "sweeten the deal" in the comfort of your own home.

MAKE YOUR OWN DATE SUGAR

Preheat oven to 400°F (200°C, or gas mark 6). Bake pitted dates for 10 minutes. Grind in food processor. Store in an airtight container at room temperature.

MAKE YOUR OWN POWDERED SUGAR

Using a high-speed blender, food processor, or coffee grinder, blend evaporated cane juice until powdery. If food processor or blender isn't efficient enough on its own, add 2 tablespoons (16 g) cornstarch to 1 cup (192 g) evaporated cane juice, and blend until powdery.

MAKE YOUR OWN MAPLE SYRUP SUBSTITUTE

If you run out of maple syrup but want to douse your pancakes with something that tastes similar, whisk up to 1 teaspoon maple extract into 1 cup (336 g) brown rice syrup or agave nectar.

MAKE YOUR OWN CHOCOLATE SAUCE OR SYRUP

Combine 1/4 cup (60 ml) liquid sweetener (such as pure maple syrup, agave nectar, or brown rice syrup) with 1/4 cup (20 g) unsweetened cocoa powder, a tiny pinch of salt, and 1 to 2 teaspoons brewed coffee, stirring until the sauce is perfectly smooth. Add flavorings, such as pure extracts or orange zest, to taste. You can replace coffee with nondairy milk or any other beverage. Store in an airtight container in fridge. Keeps well for at least a week.

MAKE YOUR OWN SUGAR SYRUPS

Combine 1 cup (192 g) Sucanat, or any healthier dry sweetener alternative, with 1/2 cup (120 ml) water in a saucepan. Bring to a boil and lower heat immediately so that it doesn't overflow. Stir and simmer for a few minutes, until crystals dissolve. The syrup will thicken as it cools. This will yield approximately 1 cup (235 ml) syrup.

If you try this with date sugar, you will need up to 2 cups (470 ml) water, and you must blend the mixture for it to be smooth. It's great on waffles and pancakes! Add flavorings such as cinnamon or vanilla and other extracts. Be sure to add extracts only after the syrup is removed from the stove, so that they don't evaporate.

This process will not work with stevia or xylitol.

MAKE YOUR OWN FRUIT PURÉES

See chapter 3 (page 63) to see how to turn dried and fresh fruit into delectable sweeteners. Swirl into unsweetened nondairy yogurts or in your breakfast bowl of oatmeal.

Sweet Recipes with Less Sugar, Healthier Sweeteners, or No Sugar at All!

A sweet tooth can be rather demanding, so satisfy it with the following recipes, which focus on sweeteners that are gentler on your glycemic index and on your general health.

PB AND (O)J WAFFLES

 Sugar sub: pure maple syrup and orange marmalade, if using all-fruit spread

 Corn Free

Packed with a mega-dose of peanut butter and no refined sugar, these are sure to keep your tummy satisfied and energy up for hours on end.

- 1 cup (256 g) natural smooth peanut butter
- 1/4 cup (60 ml) pure maple syrup
- 1 teaspoon pure orange extract
  Zest of 1 small orange
- 1/4 cup (80 g) orange marmalade
- 1 cup (80 g) quick oats
- 1 1/2 cups (180 g) whole wheat pastry flour
- 1 tablespoon (12 g) baking powder
- 1/4 teaspoon fine sea salt
- 1 1/2 cups (355 ml) nondairy milk as needed
  Nonstick cooking spray

Note: Make these into pancakes by adding 2 tablespoons to 1/4 cup (30 to 60 ml) nondairy milk to the batter. Cook as you would regular pancakes.

COMBINE PEANUT BUTTER, syrup, extract, zest, and marmalade in a large bowl.

Stir in oats, flour, baking powder, and salt. Add milk, using more if batter is too thick.

Follow waffle iron manufacturer's instructions to prepare waffles. Be sure to spray between each waffle. If your waffle maker has a thermostat, use a medium to low temperature as the nut butter makes these waffles brown a bit more quickly.

Cool waffles on a wire rack, giving them a couple of minutes to crisp up. Leftovers can be reheated in a toaster oven.

YIELD: 10 standard waffles or 12 medium pancakes

NO-KNEAD BANANA BREAD WAFFLES

 Sugar sub: pure maple syrup and banana

 Corn Free  **Soy Free**

Bread dough! Shaped like waffles! It's crispy and crusty and crazy. And best of all, it is free of refined sugars and opens the door to limitless variations.

²/3 **cup (160 ml) water, heated to lukewarm**
¹/4 **cup (64 g) natural crunchy peanut butter**
2 **tablespoons (30 ml) pure maple syrup**
3 **small (1 cup, or 240 g) ripe bananas, mashed**
2 **teaspoons active dry yeast**
1 **cup (125 g) all-purpose flour**
2 **cups (240 g) white whole wheat flour**
1 **teaspoon fine sea salt**

COMBINE WATER, peanut butter, maple syrup, and bananas; stir until combined. It doesn't matter if small chunks of peanut butter and bananas remain. Using a fork, stir in yeast until dissolved. Place flours and salt in a large bowl.

Pour liquid ingredients on top of flour mixture, and stir with a fork for a couple of minutes, until batter combines and no traces of flour are left. Gather dough in the center of the bowl with a silicone spatula.

Cover and let rise for 2 hours, or until the dough has doubled in size.

Place in the fridge for at least 8 hours or overnight. Stir dough with a fork. Grab approximately ¹/2 cup (190 g) dough, and bring it back to room temperature for about 45 minutes. (You don't have to prepare all the waffles at one time; just place the remaining dough back into the fridge and use as needed.)

Follow waffle iron manufacturer's instructions to prepare bread waffles.

Set heat level to medium low and bake for about 10 minutes or until the waffle sounds hollow when tapped and is golden brown in color.

Cool waffles on a wire rack for a few minutes before enjoying.

Serve with maple syrup, jam, vegan chocolate hazelnut spread, nondairy butter, or nut butters.

YIELD: About 6 bread waffles

PEANUT GINGER BERRY GOLDEN OATS

 Sugar sub: brown rice syrup and dried fruit (berries)

 Corn Free   Quick and Easy

 Soy Free   Wheat Free

Start with some Basic Homemade Sorta Yogurt (page 20), add fresh berries, and toss a handful of these oats on top— you've got yourself a very happy and quick breakfast or snack indeed!

2 tablespoons (30 ml) peanut oil
1/4 cup (84 g) brown rice syrup
  Pinch fine sea salt
2 teaspoons ground ginger
1/2 cup (68 g) vegan dry-roasted unsalted peanuts
2 cups (160 g) old-fashioned oats
1/2 cup (60) any dried berries

PREHEAT OVEN to 350°F (180°C, or gas mark 4).
Combine oil and syrup on a large rimmed baking sheet. Add salt, ginger, peanuts, and oats. Stir to combine, using two heat-safe silicone spatulas. Spread mixture evenly on baking sheet. Bake for 8 minutes, remove from oven to stir with spatulas, and spread evenly again. Bake for another 8 minutes or until golden brown. Stir in dried berries. Let cool on sheet to crisp. Cool completely before storing in an airtight container. Keeps for about 2 weeks.

YIELD: 2 1/2 cups (360 g)

HOMEMADE APPLESAUCE

 Sugar sub: agave nectar and fruit juice (apple)

 Corn Free   Gluten Free

 Nut Free   Soy Free

 Wheat Free

Discover how rewarding it is to make your own applesauce and use it in recipes such as Baked Apple Puddings (page 240) and Ginger Apple Coffee Cake (page 99)!

1 pound (454 g) Braeburn apples (about 4), unpeeled, quartered, cored, each quarter cut in half, and cut into 1-inch (2.5 cm) cubes
2 tablespoons (42 g) agave nectar
1 1/2 teaspoons ground cinnamon
1 tablespoon (15 ml) fresh lemon juice
1/2 cup plus 2 tablespoons (150 ml) apple juice

PREHEAT OVEN to 350°F (180°C, or gas mark 4). Combine apples, agave, cinnamon, and lemon juice in a large bowl. Place apple mixture on a baking sheet; bake for 35 minutes, carefully stirring once halfway through.
Add apple juice to the roasted apples, and blend until smooth. Add more or less juice depending on the texture of the store-bought applesauce you prefer.
Enjoy warm, at room temperature, or cold. Be sure to let cool completely before using in baking recipes.

YIELD: 2 cups (520 g)

CHOCO PB SYRUP

 Sugar sub: Sucanat

 Corn Free   Soy Free   Wheat Free

This syrup just begs to be poured over nondairy ice cream or added to milk alternatives or whatever strikes your fancy. Another win for Sucanat! Another loss for tooth decay and weight management…

1/2 cup (40 g) Dutch-processed cocoa powder
1/4 cup (64 g) creamy all-natural salted peanut butter
2 cups (384 g) Sucanat
1 cup (235 ml) water
1 teaspoon pure vanilla extract

USING A BLENDER, combine all ingredients (except vanilla) until perfectly smooth.

Transfer mixture to a medium saucepan; bring to a boil. Lower to medium heat and cook for about 3 minutes. Watch carefully and whisk occasionally, as mixture tends to rise!

Remove pan from heat and stir in vanilla.

Let cool completely. The syrup will thicken a little as it cools. Transfer to an airtight glass bottle and store in fridge. Will keep for a couple of weeks.

YIELD: 2 1/4 cups (530 ml)

Note: Fancy a hot fudge sauce—like consistency? Add an extra 2 tablespoons (10 g) Dutch-processed cocoa powder and 2 tablespoons (32 g) of the all-natural salted peanut butter.

CARAMEL BAKED APPLES

 Sugar sub: Sucanat, brown rice syrup or agave nectar, and dried fruit (raisins)

Lots and lots of caramel to cover something healthy (apples, in this case), therefore making it, well, not so healthy anymore. Hooray!

FOR CARAMEL:

- 1/4 cup (48 g) Sucanat
- 1/4 cup (56 g) nondairy butter
- 1/4 cup (84 g) brown rice syrup or dark agave nectar

FOR FILLING:

- 1 tablespoon (12 g) Sucanat
- 1 tablespoon (6 g) almond or coconut flour
- 2 tablespoons (18 g) raisins
- 1/2 teaspoon ground cinnamon
- Dash rum extract
- Dash pure almond or coconut extract
- 1 1/2 teaspoons nondairy milk
- 2 large baking apples, such as Granny Smith, cored
- 1/4 cup (40 g) dry-roasted unsalted whole almonds or (27 g) pecans, coarsely chopped

TO MAKE THE CARAMEL: Combine Sucanat, butter, and syrup or agave in medium saucepan.

Cook over medium-high heat, until butter melts. Lower heat to medium, and boil gently for 1 minute, stirring constantly. Set aside.

Preheat oven to 350°F (180°C, or gas mark 4).

TO MAKE THE FILLING: Combine Sucanat, flour, raisins, cinnamon, and extracts in a small bowl. Stir in milk until combined.

Place each apple in the center of a large piece of foil. Divide filling between apples, pushing it down into space where core used to be. Wrap foil around apple.

Place apples in 8 x 4-inch (20 x 10 cm) loaf pan. Bake for 25 minutes. Unwrap slightly, using gloves, to avoid burns.

Pour in half of the caramel, rewrap and bake for another 25 minutes.

Unwrap apples, cover with remaining caramel and chopped nuts. Bake unwrapped for another 15 minutes, or until apples are tender to the fork.

Let stand a few minutes before enjoying. Be sure to coat the apples with any caramel that slid to the bottom of the foil!

YIELD: 2 servings

REFINED-SUGAR-FREE SWEET PIZZA

 Sugar sub: pure maple syrup and banana

 Corn Free   **Soy Free**

Hand your child a slice of sweet pizza and watch for the delighted reaction! Don't mention the fact that it is not only tasty but healthy, too—it might lose some of its appeal. If you're not looking for a completely sugar-free option, how about drizzling the pizza with Choco PB Syrup (page 232) instead of jam?

> **One recipe No-Knead Banana Bread Waffles dough (page 230)**
> 1/3 **cup (85 g) natural crunchy peanut butter, heated and emulsified with 2 tablespoons (30 ml) water, or as much as needed to make it easy to spread**
> 2 **large ripe bananas, cut into 1/2-inch (1.27 cm) slices**
> 1/3 **cup (107 g) all-fruit jam, heated and emulsified with 1 tablespoon (15 ml) water if too thick, or as much water as needed to allow jam to drizzle on top**

LINE A LARGE COOKIE SHEET with parchment paper or a silicone baking mat.

Spread dough as thinly as possible without tearing; use slightly dampened hands if dough is sticky. Spread peanut butter on top. Let rest at room temperature for 1 hour. Preheat oven to 400°F (200°C, or gas mark 6). Place banana slices decoratively on top of the pizza. Bake for 20 minutes, or until the edges are golden brown.

Drizzle as much jam as you want on top. Let pizza rest for a few minutes, until cool enough to eat.

YIELD: 6 to 8 servings

CHOCOPLUM PIE

 Sugar sub: Sucanat

The sweet and spicy crust pairs up beautifully with the filling! And the pie tastes even better straight from the freezer: Give it about 15 minutes to thaw before enjoying.

FOR PLUM COMPOTE:

- 1 **cup (170 g) dried plums**
- 1 **cup (235 ml) water**
- **Scant 1¹/₂ cups (255 g) chopped bittersweet chocolate**

FOR CRUST:

- **Nonstick cooking spray**
- **Half recipe Speculoos-Spiced Cookies (page 97), unbaked and not refrigerated**

FOR FILLING:

- 12 **ounces (340 g) soft silken tofu**
- 2 **teaspoons pure vanilla extract**
- 2 **tablespoons (42 g) agave nectar**
- 1 **tablespoon (8 g) cornstarch**
- ¹/₂ **teaspoon baking powder**

TO MAKE PLUM COMPOTE: Combine plums and water in medium saucepan.
Bring to a boil. Lower heat and simmer for 15 minutes. Remove from heat. Stir in chips until melted. Using a blender, blend mixture until smooth. Let cool completely before using.

TO MAKE CRUST: Preheat oven to 350°F (180°C, or gas mark 4). Lightly coat a 9-inch (23 cm) round baking pan with cooking spray.
Crumble cookie dough evenly into pan. Press down at bottom and around lower edges of pan.
Pre-bake crust for 12 minutes or until golden brown; it will harden as it cools. Place on wire rack to cool completely.

TO MAKE FILLING: Using a blender, blend plum compote, tofu, vanilla, and agave until smooth.
Add cornstarch and baking powder; blend until combined.
Place filling in crust. Bake for 30 minutes. Turn oven off but leave pie in for another 15 minutes.
Let cool on wire rack. Transfer to freezer for 2 hours, as it is easier to slice frozen

YIELD: 8 to 10 servings

APPLE PUMPKIN PIE

 Sugar sub: agave nectar, dried fruit (dates), and fruit juice (apple)

 Soy Free

Ever wonder what would happen if an apple and a pumpkin had a pie baby together? Here's the scrumptious answer.

FOR CRUST:

1¹/2 cups (180 g) light spelt flour
 ¹/4 teaspoon fine sea salt
  2 tablespoons (42 g) agave nectar
 ¹/4 cup (60 ml) peanut oil
 ¹/4 cup (60 ml) water

FOR FILLING:

 1 cup (240 g) pitted Medjool dates, halved
2³/4 cups (650 ml) apple juice
 ³/4 cup (74 g) pecan halves
 1 tablespoon (7 g) ground cinnamon
 1 teaspoon ground ginger
 ¹/2 teaspoon ground nutmeg
 ¹/4 teaspoon fine sea salt
   Nonstick cooking spray
1¹/2 cups (366 g) pumpkin purée
 2 teaspoons pure vanilla extract
 2 teaspoons maple extract
 2 tablespoons (16 g) cornstarch

TO MAKE THE CRUST: Sift flour and salt in large bowl. Stir in remaining ingredients, adding 1 tablespoon (15 ml) of water at a time until dough forms.
Wrap dough in plastic and refrigerate for 2 hours.

TO MAKE THE FILLING: Combine dates and 3/4 cup (180 ml) of juice in a large saucepan.
Cook over medium-high heat for 6 minutes, or until dates start to dissolve.
Add pecans and spices. Cook for another minute. Add 1¹/2 cups (355 ml) juice and salt. Set aside.
Preheat oven to 375°F (190°C, or gas mark 5).
Lightly coat a 9-inch (23 cm) round cake pan with cooking spray. Sprinkle some flour on counter and roll out dough to fit pan.
Carefully transfer dough to cake pan and prick in several places with fork. Cover with parchment paper, weigh down with dry beans, and bake for 8 minutes. Remove from oven and place on wire rack.
Using a blender, blend date mixture, pumpkin, and extracts until smooth.
Combine remaining ¹/2 cup (120 ml) juice with cornstarch to make a slurry.
Add slurry to filling; blend until smooth.
Place filling into crust; bake for 50 minutes, or until center is set.
Transfer pan to wire rack. Chill overnight.

YIELD: 8 to 10 servings

CHIPOTLE CARAMEL COOKIES

 Sugar sub: Sucanat and pure maple syrup

 Corn Free

These cookies are thin and chewy and will tickle your taste buds. If the thought of adding chipotle to your sweets gives you pause, be daring and try it with the amount of chipotle called for and adjust to your liking the next time you make a batch!

- 1 cup (192 g) Sucanat
- 1/2 cup (112 g) nondairy butter
- 2 tablespoons (30 ml) pure maple syrup
- 1/2 teaspoon fine sea salt
- 1/2 teaspoon chipotle powder to taste, optional
- 2 teaspoons pure vanilla extract
- 1/2 teaspoon baking soda
- 1/2 cup (68 g) vegan dry-roasted peanuts
- 2 cups (240 g) whole wheat pastry flour

PREHEAT OVEN to 350°F (180°C, or gas mark 4). Line two large cookie sheets with parchment paper or silicone baking mats.

In a medium microwave-safe bowl, combine Sucanat, butter, syrup, salt, and chipotle.

Microwave until butter melts, sugar dissolves, and the mixture starts to boil, at least 1 minute. Stir well.

Stir in vanilla. Add baking soda, stir well; the mixture will bubble up and become a bit thicker. Stir in peanuts. Let cool for a few minutes.

Add flour and stir well.

Place 2 heaping tablespoons (55 g) dough onto prepared cookie sheets.

Flatten the cookies, as they only spread a little while baking.

Bake for 10 minutes, or until cookies start turning golden brown at the edges.

Remove from oven and let cool on sheets until firm enough to transfer to wire rack.

YIELD: 12 large cookies

PEAR CHAI ROLLS

 Sugar sub: Sucanat, fruit purée (pear), and dried fruit (pear)

A deliciously spicy roll that makes waking up in the morning a touch more bearable. The tea masala used in both the rolls and the filling is a spice mixture that you'll find in the international section of well-stocked supermarkets.

FOR ROLLS:

- 2 teaspoons active dry yeast
- $1/2$ cup (96 g) Sucanat
- $1^1/4$ cups (295 ml) nondairy milk, heated to lukewarm
- $1/4$ cup (56 g) solid coconut oil, melted
- 8 ounces (227 g) pear sauce
- $4^1/4$ cups (595 g) whole spelt flour
- 1 tablespoon (7 g) tea masala
- 2 teaspoons ground cinnamon
- 1 teaspoon ginger powder
- $1^1/2$ teaspoons sea salt

FOR FILLING:

- $1/2$ cup (120 g) pear sauce
- 2 teaspoons tea masala
- $1/4$ cup (48 g) Sucanat
- 1 cup (144 g) chopped dried pear, rehydrated in hot tea

FOR ICING:

- $1^1/2$ cups (304 g) vegan white chocolate chips
- 3 tablespoons (45 ml) nondairy milk

TO MAKE THE ROLLS: Proof yeast with Sucanat and milk for 5 minutes. Add oil and pear sauce to yeast mixture.

Combine with remaining ingredients. Stir for 4 minutes, scraping sides and gathering dough.

Cover with plastic wrap; let rise for 3 hours. Chill overnight.

Line baking sheet with parchment paper or silicone baking mat.

Grab $1/2$ cup (160 g) dough per roll. Flatten dough onto parchment paper or baking mat into thin rectangle. Spread 1 tablespoon (15 g) pear sauce over dough, leaving less than $1/2$ inch (1.27 cm) from edges. Combine masala with Sucanat in small bowl; sprinkle $1^1/2$ teaspoons on top of sauce; top with 2 tablespoons (18 g) chopped pear.

Carefully roll up dough lengthwise, using dough scraper if needed, as dough may be sticky. Pat down to form roll; repeat with remaining dough.

Cover rolls with plastic wrap; let rest for 1 hour.

Preheat oven to 375°F (190°C, or gas mark 5). Bake for 24 minutes, or until bottom of rolls sounds hollow when tapped.

Let cool on wire rack.

TO MAKE THE ICING: Combine all ingredients in small microwave-safe bowl. Heat for 1 minute in microwave; stir until chips melt.

Drizzle over rolls.

YIELD: 8 rolls

BAKED APPLE PUDDINGS

 Sugar sub: dried fruit (dates) and applesauce

 Wheat Free

A delicious baked pudding that is perfect for breakfast or dessert, and versatile enough to encourage you to incorporate other grains, nuts, and spices.

$1/2$ teaspoon nondairy butter
$1/4$ cup (20 g) old-fashioned or quick oats
$3/4$ cup (75 g) walnut halves
   4 Medjool dates, pitted and quartered
$1/2$ cup (122 g) Homemade Applesauce (page 231)
$1^1/2$ teaspoons pure vanilla extract
$1/2$ teaspoon ground cinnamon
   Pinch fine sea salt
$1/4$ teaspoon baking powder

PREHEAT OVEN to 350°F (180°C, or gas mark 4).
   Lightly coat two 6-ounce (170 g) oven-safe ramekins with butter.
   Combine oats and walnuts in food processor. Blend until finely ground.
   Add dates, and process until smooth. Add remaining ingredients and process until smooth. Divide the sticky batter into prepared ramekins. Sprinkle an extra pinch of ground cinnamon on top of each ramekin. Bake for 10 minutes, or until set. Enjoy warm, at room temperature, or cold.

YIELD: 2 servings

STICKY SWEET CRUNCHY BISCUITS

 Sugar sub: Sucanat, pure maple syrup, and dried fruit (raisins)

The preparation is slightly messy, but these crunchy biscuits are worth it!

FOR SYRUP:

- **¼ cup (48 g) Sucanat**
- **¼ cup (56 g) nondairy butter**
- **¼ cup (60 ml) pure maple syrup**
  Nonstick cooking spray

FOR FILLING:

- **⅓ cup (64 g) Sucanat**
- **2 teaspoons ground cinnamon**
- **⅓ cup (53 g) raisins**

FOR BISCUITS:

- **10 ounces (283 g) any thick jam**
- **½ cup (120 ml) nondairy milk**
- **¼ cup (60 ml) peanut oil**
- **1 tablespoon (15 ml) pure maple syrup**
- **½ teaspoon fine sea salt**
- **2 cups (280 g) whole spelt flour**
- **1 tablespoon (12 g) baking powder**
- **½ cup (40 g) old-fashioned oats**

TO MAKE THE SYRUP: Combine Sucanat, butter, and syrup in medium saucepan.
Cook over medium-high heat, until butter melts. Lower heat to medium, and cook for
1 minute, stirring constantly. Set aside.
Preheat oven to 375°F (190°C, or gas mark 5). Coat jumbo muffin pan with cooking spray.
Pour syrup into prepared muffin pan. Set aside.

TO MAKE THE FILLING: Combine all ingredients in small bowl. Set aside.

TO MAKE THE BISCUITS: Whisk jam, milk, oil, syrup, and salt in large bowl.
Sift flour and baking powder on top, stir in with oats using a fork, until just mixed.
In each muffin cup, spread scant ¼ cup (65 g) dough on top of syrup. Sprinkle 1½
tablespoons (15 g) filling on top of dough. Carefully spread remaining dough equally among
muffin cups. Place muffin pan on top of rimmed baking sheet to catch spills.
Bake for 25 minutes. Leaving oven on, remove pan to wire rack; let stand for 10 minutes.
Flip biscuits onto piece of parchment paper or silicone baking mat. Transfer to baking sheet;
bake for 10 minutes. Let cool on wire rack. Enjoy straight from fridge for crispiest results.

YIELD: 6 jumbo biscuits

SALTED CARAMEL WHEAT TREATS

 Sugar sub: organic brown sugar and brown rice syrup

 Corn Free

This snack is all about crunchy, sweet, and salty! If you can find chocolate-covered mini pretzels, more power to you. The caramel prep will melt the chocolate, but who's complaining? If not, non-chocolate-covered pretzels are just fine.

The puffed wheat can be replaced with puffed brown rice, and if you can find gluten-free pretzels, this snack is ready for the GF crowd.

6 cups (114 g) puffed wheat

8 ounces (227 g) salted mini pretzels, coarsely chopped

1/4 cup (56 g) nondairy butter

1/2 cup (96 g) organic brown sugar 🔄

1/2 cup (168 g) brown rice syrup 🔄

1/4 teaspoon baking soda

1 cup (137 g) vegan dry-roasted salted peanuts

1 teaspoon pure vanilla extract

LINE A 9-inch (23 cm) square pan with waxed paper.

Combine wheat and pretzels in a large bowl.

In a medium saucepan, combine butter, sugar, and syrup. Bring to a boil and cook for 1 minute, stirring constantly.

Remove from heat. Stir in baking soda; the caramel will thicken up. Stir in peanuts and vanilla. Pour mixture over wheat and pretzels; stir until combined.

Spread mixture in pan and press down with another piece of waxed paper. Store in fridge for 1 hour until set before slicing and enjoying.

YIELD: 12 treats or more

NO-SUGAR-ADDED CAKEY CHOCOLATE COOKIES

 Sugar sub: xylitol, stevia, or blends

We're happy to report that these cookies have no added sugar! And their texture is perfect for whoopie pies! Sandwich them around ABC Mousse Topper (page 244) for an indulgent treat.

2¹/2 cups (313 g) all-purpose flour
 ¹/2 cup (40 g) sifted unsweetened cocoa powder
   1 tablespoon (3 g) instant espresso powder, optional
1¹/2 teaspoons baking powder
 ¹/2 teaspoon baking soda
 ¹/4 teaspoon salt
 ¹/4 teaspoon ground cinnamon
   1 cup (192 g) xylitol (or stevia and xylitol blend, such as Truvia or Z-Sweet)
1¹/4 cups (295 ml) nondairy milk
 ¹/4 cup (60 ml) oil
   1 tablespoon (15 ml) pure vanilla extract
   6 ounces (170 g) nondairy yogurt

PREHEAT OVEN to 350°F (180°C, or gas mark 4). Line baking sheet with parchment paper or silicone baking mat.

In a large mixing bowl, combine flour, cocoa powder, espresso powder, baking powder, baking soda, salt, cinnamon, and xylitol.

In a separate bowl, combine milk, oil, vanilla, and yogurt.

Fold wet ingredients into dry, being careful not to overmix.

Spoon 2 tablespoons (30 g) batter onto baking sheet, allowing for room to spread, about 6 cookies per sheet.

Bake for 12 to 15 minutes, until cookies are crackly on top. Cool briefly before transferring to wire rack to cool completely.

YIELD: 24 cookies

ABC MOUSSE TOPPER

 Sugar sub: fruit (banana) and sugar contained in chocolate

Corn Free   **Wheat Free**

Sweetened with fruit and no added refined sugar (other than what's in the chocolate), this tasty topper is healthier than standard frostings that are packed with shortening and powdered sugar.

1^1/$_2$ cups (264 g) bittersweet chocolate chunks
^{1}/$_4$ cup (60 ml) coconut milk
2 large ripe bananas, sliced (1^1/$_2$ cups, or 240 g)
2 small ripe Haas avocados, pitted and peeled
1 teaspoon pure vanilla extract
^{1}/$_2$ teaspoon rum extract
^{1}/$_2$ teaspoon pure orange extract
^{1}/$_4$ teaspoon pure coconut extract
   Pinch sea salt

MELT CHOCOLATE in microwave. Place milk, bananas, avocados, extracts, and salt in food processor. Blend until smooth. Add half the melted chocolate; process until smooth, scraping sides occasionally. Transfer into large bowl or individual dessert dishes. Swirl what's left of the melted chocolate with a toothpick, making chocolate ribbons that will harden when chilled, providing a crunchy surprise when the mousse is enjoyed. Chill before serving.

YIELD: 3^1/$_2$ cups (963 g), enough for 6 desserts or to frost 24 cupcakes, 12 whoopie pies, or 1 double-layer cake

LEMON RASPBERRY FRUIT-FILLED DESSERT

 Sugar sub: just naturally sugar free!

 Gluten Free   Nut Free

 Quick and Easy   Soy Free

 Wheat Free

This dessert is tart and light and perfect on a hot afternoon when you can't be bothered to pull out the ice cream maker!

1 ripe banana

1/4 cup (60 ml) fresh lemon juice

1 cup (123 g) fresh or frozen raspberries

2 cups (470 ml) rice or light coconut milk

1 teaspoon gluten-free pure vanilla extract

3 tablespoons (36 g) Z-Sweet, or 1 tablespoon (12 g) Truvia or other stevia blend, or 1/4 teaspoon stevia liquid or pure stevia powder

1 teaspoon xanthan gum

ADD ALL INGREDIENTS to blender and blend until smooth and thickened.
Chill in the fridge before serving.

YIELD: 3 cups (750 ml)

Note: This mixture works well for making ice pops. Just pour the blended ingredients into ice-pop molds, freeze, and enjoy a refreshing no-sugar-added treat!

BANANA CHUTNEY

 Sugar sub: fruit (banana) and Sucanat

 Corn Free   Soy Free

 Wheat Free

This deliciously sweet and tangy chutney tastes great served with roasted sweet potatoes or anywhere chutneys are called for.

It is also the star of our Banana Chutney Quick Bread (page 82).

1 tablespoon (15 ml) sesame oil (not toasted)

1/4 cup (40 g) chopped red onion

2 medium bananas, cut in large chunks

1 teaspoon coarse sea salt

4 ounces (112 g) fire-roasted diced green chilies

Pinch red pepper flakes, optional

1/2 cup (96 g) Sucanat

1/2 cup (120 ml) apple cider vinegar

1 teaspoon garam masala

1 teaspoon ground ginger

1 teaspoon cumin seeds

2 teaspoons tamarind paste

COMBINE OIL and red onion in a medium saucepan. Cook over medium-high heat for 2 minutes, until the onion turns translucent. Add remaining ingredients, and cook over medium heat for 10 minutes, until thickened.
Place mixture in mason jars and close tightly while still hot. Store in fridge. Keeps well for at least 1 week.

YIELD: 2 cups (480 g)

Chapter 9
For Better-, Lower-, and No-Fat Foods: HOW TO SUBSTITUTE FOR FAT

LET'S FACE IT: Fat adds flavor. But cutting back and choosing healthier, more natural alternatives doesn't have to mean you've seen the end of great tasting meals.

Consider the Facts

Good fats, bad fats, trans fats, saturated fats, unsaturated fats, monounsaturated fats, and hydrogenated fats . . . the list goes on and on. Truth be told, not all fats are bad for us. In fact, essential fatty acids DHA and EPA are essential for proper brain and nerve function.

But not all fats are created equal. Naturally occurring fats found in nuts, avocados, seeds, grains, and other plant foods are the best way to maintain a well-balanced diet.

It's when we start adding fats to our foods that we need to be careful, especially when purchasing ready-made foods.

Take for example hydrogenated fat, a common added fat found in processed prepackaged foods. The process of hydrogenating fats is basically adding hydrogen to liquid oil, which keeps it solid at room temperature and increases the shelf life and melting point. There is absolutely nothing natural or nutritious about it. Studies show a link between these artery-clogging fats and an increased risk of cancer, heart disease, and obesity, and a compromised immune system.

Moderation is key, and there are so many ways of incorporating healthier fats or switching to flavorful lower fat options that it'd be a pity for anyone to believe the only way to maintain a healthy weight and body is to give up on taste altogether.

Let's get acquainted with a few substitutions by looking at the following chart.

Fat Substitution Guidelines

IF THE ORIGINAL RECIPE CALLS FOR...	REPLACE WITH...
½ cup (120 ml) melted dairy or nondairy butter	• ½ cup (120 ml) grapeseed, olive, peanut, or melted coconut oil (depending on use, choose mild-flavored oils)
½ cup (112 g) dairy butter	• ½ cup (112 g) nonhydrogenated, nondairy butter
½ cup (112 g) dairy butter (in cakes, muffins, waffles, cookies) and you want to switch to using less fat	• ¼ cup (56 g) nondairy butter plus ¼ cup (60 g) unsweetened applesauce or other mild-flavored fruit or vegetable purée (dried plums, avocado, beans). Note: You can reduce the amount of fat to only 2 tablespoons (28 g) nondairy butter while increasing the amount of applesauce or other fruit purée accordingly but not lower than that as it could make for gummy baked goods. Using fruit purée in cookies to replace fat will make for cakey cookies.
½ cup (112 g) dairy butter (in cakes, muffins, waffles) and you want to switch to using healthier, less heavily processed source of fat than nondairy butter	• ¼ cup plus 3 tablespoons (105 ml) vegetable or other mild-flavored oil
½ cup (112 g) dairy butter (in cookies) and you want to switch to using healthier, less heavily processed source of fat than nondairy butter	• ¼ cup (60 ml) vegetable or other mild-flavored oil (may change the texture of cookies)
½ cup (112 g) shortening	• ½ cup (112 g) nonhydrogenated vegetable shortening
½ cup (120 ml) vegetable oil (in cakes, muffins, waffles, cookies), and you want to switch to using less fat	• ¼ cup (60 ml) oil plus ¼ cup (60 g) unsweetened applesauce or other mild-flavored fruit or vegetable purée (avocado, beans). Note: You can reduce the amount of fat to only 2 tablespoons (30 ml) oil while increasing the amount of applesauce or other fruit purée accordingly but not lower than that as it could make for gummy baked goods. Using fruit purée in cookies to replace fat will make for cakey cookies.

(continued on next page)

(continued from previous page)

IF THE ORIGINAL RECIPE CALLS FOR...	REPLACE WITH...
Vegetable oil (to sauté vegetables), and you want to reduce or suppress fat	• Add a pinch salt to release moisture from veggies, plus vegetable broth, water, wine, or beer as needed • Carefully coat warm pan away from the stove with nonstick cooking spray, or use a spray filled with oil to minimize the amount needed • Fresh herbs and seasonings, to boost flavor

VEGANIZED!: SAMPLE RECIPE

Let's have a look at the following traditional recipe for an example of how we would

WACKY CAKE

This recipe has been adapted from the well-known Depression-era egg-free and dairy-free cake that has pleased palates for ages.

Nonstick cooking spray

1¹/2 cups (185 g) all-purpose flour

> We would use 1 cup (120 g) whole wheat pastry flour combined with ¹/2 cup (63 g) all-purpose flour.

1 cup (192 g) white sugar

> We would use 1 cup (120 g) powdered evaporated cane juice.

¹/4 cup (20 g) unsweetened **cocoa powder**

1 teaspoon baking soda

> We would add an optional ¹/3 cup (59 g) nondairy semisweet chocolate chips for extra chocolate goodness.

¹/2 teaspoon salt

1 teaspoon pure vanilla extract

1 tablespoon (15 ml) apple cider vinegar

¹/4 cup plus 1 tablespoon (75 ml) vegetable oil

> We would use ¹/4 cup plus 1 tablespoon (76 g) unsweetened applesauce.

1 cup (235 ml) water

> We would replace this with 1 cup (235 ml) nondairy milk to compensate for the fact no oil is used.

PREHEAT OVEN to 350°F (180°C, or gas mark 4). Lightly coat an 8-inch (20 cm) square baking pan with cooking spray.

In a large bowl, sift together flours, powdered evaporated cane juice, cocoa powder, baking soda, and salt. Set aside.

In a small bowl, whisk together vanilla, vinegar, applesauce, and milk.

Fold wet ingredients into dry, being careful not to overmix. Fold in chocolate chips, if using.

Place batter into prepared pan. Bake for 30 to 40 minutes, or until a toothpick inserted in center comes out clean. Place on a wire rack and cool completely.

YIELD: 10 servings

Finding Fat Alternatives and Substitutes at the Store

Most supermarkets and health-food stores carry a large variety of oils, nondairy butters, vegetable broths, and juices, as well as:

Applesauce: See page 231 for recipe. Great as a substitute for fat in baked goods like cakes, muffins, and cookies (if the latter are meant to be a bit cakier). Acts as an egg substitute at the same time. Other fruit and vegetable purées such as pumpkin, dried plum, and sweet potatoes work just as well, as long as you pair them according to flavor so that they don't overwhelm the dish.

Borage oil: Great for salads or other foods that don't require heating.

Canola oil: Also known as rapeseed, best when purchased organic or non-GMO. Very neutral in flavor. Great for baking and sautés.

Corn oil: Great for sauces, foods cooked at low heat, frying, and for all-purpose cooking.

Coconut oil: Adds great flavor. Great for baking, sauces, sautés, and foods cooked at low heat.

Flaxseed oil: Rich in omega-3 fatty acids. Great for salads and foods that won't be heated.

Fresh and **dried herbs:** A great way to add flavor with less or no fat while decreasing the need for extra salt.

Grapeseed oil: Great for salads, baking, and sautés.

Hemp oil: Rich in essential omegas. Great for sauces and foods cooked at low heat.

Nondairy butter: Purchase the nonhydrogenated kind. Use in moderation as it is still a highly processed product.

Olive oil: Extra-virgin is great for sauces, drizzling, and foods cooked at low heat. Regular is great for baking, sautés, frying, and for all-purpose cooking.

Peanut oil: Great for sauces, foods cooked at low heat, frying, and for all-purpose cooking.

Safflower oil: Great for salads and other foods that won't be heated.

Sesame oil: Toasted variety delivers bold flavor and is great for sauces and foods cooked at low heat. Regular is milder in flavor, great for frying and all-purpose cooking.

Walnut oil: Rich in essential omegas. Great for salads and other foods that won't be heated.

Vegetable broth, white and **red wine, beer, fruit** and **vegetable Juices:** Perfect for cooking with less or no fat, these add flavor while retaining food's texture and moisture. Choose low-sodium vegetable broth and vegetable juices, or make your own.

Vegetable oil: Mild-flavored. Great for all-purpose cooking and baking.

Making Fat Alternatives and Substitutes at Home

It's surprisingly easy to make healthy and lower-fat alternatives to oil and butter at home.

USE FRESH HERBS, SPICES, VEGETABLE BROTH, VEGETABLE AND FRUIT JUICE, WHITE AND RED WINE, VINEGARS, BEER

Make use of plenty of fresh herbs, vegetable broth, white wine, and other seasonings in your cooking. It's a well-known fact that fat adds flavor to foods, but there are other ways to contribute flavor without adding as many calories or as much fat. Add just a small amount of healthy fat to make sure foods don't stick, and then turn to fresh herbs to add a burst of freshness and vitamins, along with a pleasing visual aspect that takes any dish from boring to extraordinary. If fresh herbs are hard to obtain depending on the season, dried herbs are a great substitution. Usually 1 teaspoon of a dried herb will provide the same amount of flavor as 1 tablespoon (3 g, depending on the herb) of its fresh counterpart.

Using broth, wine, vinegars, and beer (pick nonalcoholic versions if you prefer) is another great way to impart flavor to your food, although it won't be as low in calories as using broth or herbs. There are so many great flavors available at your market; you won't know which to select! Check www.barnivore.com to see which wines and beers are vegan-friendly.

These liquids can both deglaze a pan and steam your veggies without adding an outrageous amount of fat. Choose low-sodium broths in order to control your salt intake.

USE FRUIT OR VEGETABLE PURÉES

Replacing fat with fruit purées in baked goods is a healthy and flavorful substitution that also reduces the amount of sugar, depending on the sweetness of the purée in use. Be careful which fruit you pick—some, like bananas, have strong flavors that can take over the whole dish. Get creative and test out avocado purée, pumpkin purée and even beans! Note that you can reduce the amount of fat to only 2 tablespoons (30 ml) oil or (28 g) nondairy butter in baked goods while increasing the amount of applesauce or other fruit purée accordingly—but not lower than that as it could make for gummy treats. Also, using fruit purée in cookies to replace fat will change the texture, making for cakey cookies.

USE OLIVE OIL BUTTER

See chapter 1 (page 19) for recipe!

USE NONSTICK COOKING SPRAY

A great way to keep the amount of fat and calories down while making sure that your baked goods won't disappoint by sticking to the pan and breaking into a million pieces. While on the subject of sprays, consider filling a small spray bottle with any oil in order to keep the amounts used to a minimum.

USE NUT BUTTERS

As a general rule, 1 cup (about 128 g, depending on the nut) nuts turns into $1/2$ cup (128 g) nut butter. The higher in fat the nuts, the smoother the butter. If you want crunchy butter, add a handful of extra nuts when you're almost done with the processing.

Make your nut butters easier to spread and lower in fat by emulsifying with just a little water, fruit juice, liquid sweetener, oil, or nondairy milk.

Ever think about a flavored nut butter? Think cinnamon and raisin, maple syrup, or even melted chocolate!

We find that toasting nuts makes them easier to turn into butter. Just spread them on a baking sheet and bake at 400°F (200°C, or gas mark 6) for 5 to 10 minutes, until fragrant.

Make small batches of homemade nut butters, as they are more perishable than store-bought, and store in the fridge for up to a month. Store your nuts in the fridge or freezer so that they don't turn rancid due to their high fat content.

While using a Vita-Mix makes preparing nut butters a breeze, it is an expensive investment that neither we nor a lot of people can afford, and you can rely on an efficient food processor to do the job. It will take more time for them to release their oil, up to 15 minutes depending on the nuts and machine. Take occasional breaks so that the motor doesn't overheat on you!

Recipes with Less Fat, Healthier Fat, or No Fat at All!

If you are looking for ways to reduce your fat intake, or you simply want to switch to fats that are better for your heart and general well-being, the following recipes will help you take a step towards happier arteries.

WHITE CHOCOLATE ORANGE CORN MUFFINS

 Fat sub: fruit purée (apple)

 Low Fat

Applesauce provides moisture so that no added fat is necessary, which makes this a tasty and healthy muffin to get your day started right.

- 1/2 cup (168 g) agave nectar
- 1/2 cup (120 ml) nondairy milk
- 1/2 teaspoon orange zest
- 1/2 cup (120 ml) fresh orange juice
- 2 teaspoons apple cider vinegar
- 1/4 cup (60 g) unsweetened applesauce %
- 1/2 cup (102 g) vegan white chocolate chips or White Chocolate (page 23), cut into chunks
- 1/2 teaspoon fine sea salt
- 1 1/4 cups (150 g) whole wheat pastry flour
- 1 1/4 cups (175 g) cornmeal
- 1 teaspoon baking soda

PREHEAT OVEN to 350°F (180°C, or gas mark 4). Line a standard muffin tin with paper liners.
   Combine agave, milk, zest, juice, vinegar, applesauce, chocolate chips, and salt in a large bowl.
   Add flour, cornmeal, and baking soda; stir until combined.
   Divide batter into prepared muffin tin, filling almost to the top of paper liner.
   Bake for 18 minutes, or until golden brown on top and firm when touched.
   Let cool on wire rack.

YIELD: 12 muffins

CHOCOLATE MUFFINS WITH FRUITY CHOCOLATE TOPPER

 Fat sub: dried fruit purée (prunes) made with fruit juice (apple)

 Corn Free   Low Fat   Soy Free

Tired of low-fat treats that are as dry as can be and low in flavor to boot?

Give these tender and delicious muffins a try, and rest assured the dried plums won't overwhelm the flavor, but they will throw extra fiber and vitamins into the mix.

FOR TOPPER:
- 1/2 cup (160 g) all-fruit spread
- 1/2 cup (40 g) unsweetened cocoa powder

FOR MUFFINS:
- 2 cups (470 ml) spiced apple cider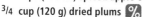
- 3/4 cup (120 g) dried plums
- 1/4 teaspoon fine sea salt
- 1 tablespoon (15 ml) pure vanilla extract
- 1 cup (192 g) raw sugar
- 1 cup (80 g) unsweetened cocoa powder
- 2 teaspoons apple cider vinegar
- 2/3 cup (100 g) chopped semisweet chocolate
- 2 1/2 cups (300 g) whole wheat pastry flour
- 2 teaspoons baking soda

TO MAKE THE TOPPER: Combine fruit spread and cocoa powder in a bowl. The mixture should be easy to spread. Set aside.

TO MAKE THE MUFFINS: Bring cider to a boil. Add plums and simmer for 8 minutes. Transfer to blender along with salt, vanilla, sugar, cocoa powder, and vinegar.

Blend until smooth. Set aside to cool completely before using.

Preheat oven to 350°F (180°C, or gas mark 4). Line 2 standard muffin tins with 16 paper liners.

Place wet ingredients in a large bowl. Add chopped chocolate. Sift flour and baking soda on top.

Fold dry ingredients into wet, being careful not to overmix. If batter is too dry, add extra apple cider, 1 tablespoon (15 ml) at a time, stirring just to combine.

Place batter into prepared muffin tins, reaching the top of the liner.

Bake for 18 minutes, or until a toothpick inserted in the center comes out clean. Let cool completely before applying topper.

YIELD: 16 muffins

TOMATO ALMOND HUMMUS

 Fat sub: natural fats in olive oil and nut butter (almond)

 Corn Free   Quick and Easy

 Wheat Free

Use this as you would any hummus: on bread or chips, or on top of roasted potatoes or veggies. By thinning it out with a little excess water from rehydrating the sun-dried tomatoes, you've got a rich and wonderful sauce for pasta.

- 3 medium Roma tomatoes, halved
- 1 large clove garlic
- 3 tablespoons (45 ml) olive oil
- 3 tablespoons (45 ml) red wine vinegar
- 2 teaspoons smoked paprika
- $1/2$ teaspoon ground black pepper
- $1/4$ cup (64 g) toasted almond butter
- $1/2$ teaspoon fine sea salt to taste
- 6 sun-dried tomatoes, rehydrated in $1/4$ cup (60 ml) boiling water for 10 minutes; squeeze out excess water before using
- 1 can (15 ounces, or 425 g) chickpeas, drained and rinsed

COMBINE ALL INGREDIENTS in blender or food processor. Blend until smooth.
Serve at room temperature or chilled. Store in airtight container in fridge.

YIELD: 3 cups (650 g)

FAT-FREE PLUM DIPPING SAUCE

 Fat sub: naturally fat free with enhanced flavors!

 Corn Free   Low Fat

 Nut Free   Soy Free

 Quick and Easy

Great to enjoy with Crispy Tofu Veggie Spring Rolls (page 124) or anything that needs a kick without extra fat. Add some chopped salted peanuts if you fancy, for some crunch.

- 2 2.5-ounce (71 g) jars organic plum baby food
- 3 tablespoons to $1/4$ cup (45 to 60 ml) seasoned rice vinegar
  Pinch fine sea salt
- 1 teaspoon ground ginger
- 2 teaspoons to 1 tablespoon (15 ml) sambal oelek (a spicy condiment available in the international food section of well-stocked supermarkets)
- 3 tablespoons (63 g) brown rice syrup or agave nectar
- 1 clove garlic, peeled and minced
- 1 scallion, chopped

COMBINE ALL INGREDIENTS in a medium bowl. Store in an airtight container in the fridge. Keeps well for about a week.

YIELD: 1 cup (235 ml)

ROASTED SQUASH AND LENTIL SPREAD

 Fat sub: natural fats in olive oil and seed butter (sesame)

 Gluten Free **Soy Free** **Wheat Free**

This spread is great in sandwiches and as a "BFF" to your roasted potatoes or steamed veggies. You can also thin it out a little and use as a sauce for pasta. Or be unladylike like us and eat it as is with a spoon.

2³/4 cups (564 g) butternut squash cubes

2 tablespoons (30 ml) extra-virgin olive oil

1 cup (192 g) uncooked red lentils, cooked in 1 quart (940 ml) water for 20 minutes or until mushy, drained

1 tablespoon (6 g) garam masala

¹/4 cup (64 g) tahini or any nut butter

1 clove garlic, peeled and chopped

1 tablespoon (7 g) granulated onion

1 teaspoon ground coriander

¹/2 teaspoon cayenne pepper to taste

1 teaspoon paprika

1 teaspoon fine sea salt

³/4 cup (180 ml) unsweetened almond milk, or vegetable broth for less salt

TO ROAST SQUASH CUBES preheat oven to 400°F (200°C, or gas mark 6). Toss cubes with olive oil and spread on baking sheet. Roast for 45 minutes, or until tender.

Combine all ingredients (except milk or broth) in food processor or blender.

Blend until smooth. Add milk or broth as needed, depending on desired thickness. Enjoy warm, at room temperature, or cold.

YIELD: 4 cups (1 kg)

TAMARIND ALMOND DIP

 Fat sub: natural fats in nut butter (almond)

 Corn Free   **Quick and Easy**

Soy Free

This flavorful dip is just right served with Crispy Tofu Veggie Spring Rolls (page 124).

1/4 cup (60 ml) filtered water
1/4 cup (64 g) crunchy natural almond butter
  3 tablespoons (45 ml) seasoned rice vinegar
  2 tablespoons (20 g) tamarind paste
     Squirt sriracha to taste
  2 tablespoons (42 g) agave nectar
  1 teaspoon ground ginger
  1 clove garlic, peeled and chopped

COMBINE ALL INGREDIENTS in food processor or blender, and blend until smooth. Store in an airtight container in the fridge. Keeps well for about a week.

YIELD: 1 cup (235 g)

NO-FAT BALSAMIC MUSTARD DRESSING

 Fat sub: no oil added

 Corn Free   **Low Fat**

Nut Free   **Quick and Easy**

Wait! No oil? That's right. A super kicky dressing without an ounce of fat! Enjoy over dark leafy greens or warmed with steamed kale.

1/2 cup (120 g) mild Dijon mustard
1/2 cup (120 ml) balsamic vinegar
  2 tablespoons (30 ml) agave nectar
  1 tablespoon (8 g) garlic powder
  1 tablespoon (8 g) onion powder
1/4 teaspoon salt
     Black pepper to taste

COMBINE ALL INGREDIENTS in an airtight container, cover with lid, and shake to mix. Store in the fridge until ready to use.

YIELD: Just over 1 cup (270 ml)

SPICY MANGO CUCUMBER SALAD

 Fat sub: natural fats in nut butter (peanut) and sesame oil

 Corn Free   **Wheat Free**

A refreshing salad that makes for a great summer meal, especially when accompanied by Yucca Fries (page 178) or toasted baguette slices. For best results, combine cucumber slices with 1 teaspoon fine sea salt, and leave to sweat for 1 hour, rinsing off salt before stirring into salad.

$^1/_4$ cup (30 g) chopped chili-coated dried mango (or add $^1/_2$ teaspoon smoked paprika and $^1/_8$ teaspoon cayenne pepper to sweetened dried mango)
2 tablespoons (32 g) crunchy natural peanut butter
1 clove garlic, peeled and minced
2 scallions, chopped
    Squirt sriracha to taste
2 teaspoons brown rice syrup
$^1/_4$ cup (60 ml) rice vinegar
    Pinch fine sea salt to taste
2 teaspoons toasted sesame oil
1 English cucumber, thinly sliced
4 celery stalks, thinly sliced
    Chopped fresh cilantro, for garnish

COMBINE FIRST 9 ingredients in a large bowl.
    Add cucumber and celery, stir to combine. Sprinkle with fresh cilantro.

YIELD: 2 servings

KABLOOEY!

 Fat sub: just naturally low in fat!

 Corn Free   **Quick and Easy**

 Soy Free

This spin on classic tabouli, made with kasha here, tastes great served warm or cold. Though not really low in fat, it contains just a small amount of heart-healthy olive oil to add to the flavor. If you cannot find toasted kasha, simply toast your groats in a dry pan before boiling.

2 cups (470 g) vegetable broth
1 cup (168 g) toasted kasha (or buckwheat groats)
1 cup (60 g) chopped fresh parsley
8 sun-dried tomatoes, packed in oil, chopped
1 tablespoon (15 g) minced garlic
2 tablespoons (30 ml) olive oil
$^1/_4$ cup (60 ml) fresh lemon juice
    Salt and pepper to taste

BRING BROTH to boil.
    Add groats and simmer until liquid is absorbed, between 5 to 15 minutes (cooking time varies depending on brand). Stir occasionally to prevent scorching on the bottom of the pot. Add extra liquid if needed.
    Remove from heat.
    Stir in remaining ingredients.
    Serve warm or cold.

YIELD: 2 main-dish or 4 side-dish servings

JALAPENO CRANBERRY PAPAYA COUSCOUS

 Fat sub: no oil added

 Corn Free Low Fat

Sweet, spicy, and a little bit sour, with no added fat. Tastes scrumptious hot or cold. We prefer to eat it cold over a bed of baby spinach or steamed kale.

Skip the optional tofu or tempeh, and you've made the dish soy free!

- 2 jalapeños, seeded and cored
- 2 cups (470 ml) water
- 1 cup (173 g) uncooked pearl couscous (Israeli couscous)
- 6 ounces (170 g) extra-firm tofu or tempeh, finely chopped (optional)
- 1/2 papaya, finely diced
- 1/4 cup (30 g) dried cranberries
- 1/4 cup (64 g) sunflower seeds
- 2 tablespoons (30 ml) fresh lemon or lime juice
- 2 teaspoons rice vinegar
- 1 teaspoon agave nectar
- 1/2 teaspoon ground cumin
- 1/2 teaspoon chili powder
  Salt and pepper to taste

PREHEAT OVEN to 350°F (180°C, or gas mark 4). Roast jalapeños in oven for 30 minutes.

While jalapeños are roasting, prepare couscous by bringing 2 cups water to a boil. Add couscous, stir, remove from heat, cover, and let sit for about 20 minutes to absorb liquid. Fluff with a fork.

Once jalapeños are roasted and cool enough to touch, finely dice them.

In a large mixing bowl, combine remaining ingredients with jalapeños and couscous.

YIELD: 8 side-dish servings

SMASHED PEPPER PEAS

 Fat sub: just naturally low in fat!

 Corn Free Gluten Free

 Low Fat Nut Free

 Quick and Easy Soy Free

 Wheat Free

Peas are good, but these peas are awesome. Serve as an alternative to mashed potatoes. . . without any added fat!

- 3/4 cup (180 ml) vegetable broth
- 1 medium yellow onion, diced
- 2 cloves garlic, peeled and minced
- 2 tablespoons (30 g) prepared horseradish
- 2 tablespoons (30 g) mild Dijon mustard
- 1 to 2 teaspoons ground black pepper to taste
- 1 pound (454 g) fresh or frozen peas, thawed
  Salt to taste

BRING BROTH to a boil. Reduce to a simmer.

Add onion and garlic, cover, and cook over medium-low heat for 20 minutes, or until vegetables are soft and most of the liquid has been absorbed.

Stir in horseradish, mustard, pepper, and peas. Heat through.

Using a blender or food processor, purée until mostly smooth but still a little chunky. Add salt to taste.

YIELD: 4 to 6 servings

COLD MULTIGRAIN SALAD

 Fat sub: no oil added

 Corn Free   **Low Fat**

 Nut Free

This recipe yields a huge amount of a delicious salad that is perfect for potlucks.

FOR SALAD:

- 1 cup (200 g) uncooked pearl barley
- 1 cup (168 g) uncooked red or white quinoa
- 1 cup (180 g) uncooked wild rice blend
- 1/3 cup (28 g) finely chopped green onion
- 1 shallot, diced
- 1 red or green bell pepper, cored, seeded, and diced
- 2 tablespoons (30 g) minced garlic
- 1 cup (60 g) packed, finely chopped fresh parsley

FOR DRESSING:

- 1/4 cup (60 ml) apple cider vinegar
- 1/4 cup (60 ml) Bragg Liquid Aminos or soy sauce
- 1/4 cup (84 g) agave nectar
- 1 tablespoon (4 g) red pepper flakes to taste
- Salt and pepper to taste

TO MAKE THE SALAD: To cook barley, the ratio of grain to water is 1: 3. To cook rice and quinoa, the ratio is 1:2. It is not recommended to cook the grains together. You can, however, cook them in the oven at the same time. Mix the proper ratios, each into a separate covered oven-safe casserole, and bake at 350°F (180°C, or gas mark 4) until water is absorbed and grains are tender, about 45 minutes to 1 hour. Fluff with a fork.

Refrigerate grains until ready to use.

Once grains are chilled, mix them together in a large bowl. Add green onion, shallot, bell pepper, garlic, and parsley.

To make the dressing: Whisk together vinegar, Bragg's or soy sauce, agave, red pepper flakes, salt, and pepper in small bowl.

Add dressing to salad; toss to mix.

Keep refrigerated until ready to serve.

YIELD: 24 side-dish servings

ORANGE MOLASSES GRILLED TEMPEH KEBABS IN CURLY MUSTARD GREENS

 Fat sub: coconut oil in curly mustard + no oil added in tempeh

 Corn Free   **Wheat Free**

Sweet, savory, smoky, and spicy: It all works together so well here to make this meal a real showstopper. Remember to soak your skewers in water for thirty minutes before using them to thread the tempeh.

FOR TEMPEH:

$1^1/_2$ cups (355 ml) orange juice

$^1/_4$ cup (60 ml) apple cider vinegar

2 tablespoons (44 g) blackstrap molasses

1 tablespoon (15 g) whole-grain mustard

1 pound (454 g) tempeh, cut into bite-size chunks and placed on skewers (6 to 8 pieces per skewer)

FOR CURLY MUSTARD:

2 tablespoons (28 g) solid coconut oil

1 pound (454 g) curly mustard greens, trimmed and chopped

1 teaspoon blended chipotle and adobo sauce

2 cloves garlic, peeled and minced

$^1/_3$ cup (53 g) chopped onion

1 teaspoon liquid smoke

TO MAKE THE TEMPEH: Add juice, vinegar, and molasses to a small pot and bring to a boil.
Reduce to medium heat and cook for 20 minutes, or until reduced to $^1/_2$ cup (120 ml).
Remove from heat and whisk in mustard.
Coat skewers with sauce.
Using a barbecue or grill pan, over high heat, cook skewers for 5 minutes per side, or until heated through, and grill marks appear.

TO MAKE THE CURLY MUSTARD: Combine all ingredients in a large saucepan.
Cook over medium-high heat until curly mustard is wilted, about 6 minutes, stirring occasionally.
Lower heat to a simmer and cover. Cook for another 8 minutes, or until tougher parts of curly mustard are tender.
Serve alongside tempeh skewers.

YIELD: 4 servings

GRAHAM WANNABE QUICK BREAD

 **Fat sub:** peanut oil

When Celine's husband first tried a bite of this bread, he said that its flavor was reminiscent of graham crackers, hence the name.

Enjoy this lightly sweetened bread with just a touch of nondairy butter.

Nonstick cooking spray
$1^1/2$ **cups (355 ml) soymilk**
2 **teaspoons apple cider vinegar**
$1/4$ **cup plus 2 tablespoons (60 g) raisins, optional**
$1/4$ **cup (88 g) regular molasses**
$1/2$ **cup (120 ml) pure maple syrup**
$1/2$ **teaspoon fine sea salt**
2 **tablespoons (30 ml) peanut oil**
1 **teaspoon pure vanilla extract**
3 **cups (360 g) whole wheat pastry flour**
1 **teaspoon ground cinnamon**
2 **teaspoons baking powder**
$1/2$ **teaspoon baking soda**

PREHEAT OVEN to 350°F (180°C, or gas mark 4). Lightly coat a 9 x 5-inch (23 x 13 cm) loaf pan with cooking spray.

Combine vinegar and soymilk in a large bowl; it will curdle and become like buttermilk.

Whisk raisins, if using, together with molasses, maple syrup, salt, oil, and vanilla into buttermilk mixture.

Add flour in three batches, whisking in between each batch, just to incorporate and being careful not to overmix. Stir in cinnamon, baking powder, and baking soda with the last batch of flour.

Pour batter into prepared pan.

Bake for 45 to 50 minutes, or until a toothpick inserted in the center comes out clean.

Transfer bread to a wire rack and let cool completely before slicing or storing.

YIELD: One 9-inch (23 cm) loaf

CAROB TREATS

 Fat sub: natural fats in nuts (walnuts)

 Corn Free   **Quick and Easy**

How about a simple dessert or breakfast treat that packs a nutritional punch, contains omega-3s (compliments of the walnuts), and is a distant cousin of pudding—and it comes with an enticingly chewy top?

- 1/2 teaspoon nondairy butter
- 3/4 cup (75 g) walnut halves
- 1/4 cup (48 g) raw sugar
- 1/4 cup (20 g) old-fashioned or quick oats
- 1 medium-size, ripe banana
- 1 1/2 teaspoons pure vanilla extract
  Pinch fine sea salt
- 1/4 cup (26 g) carob powder
- 1/4 teaspoon baking powder

PREHEAT OVEN to 350°F (180°C, or gas mark 4). Lightly coat two 6-ounce (170-g) oven-safe ramekins with butter.

Combine walnuts, sugar, and oats in food processor. Blend until finely ground. Add remaining ingredients and process until smooth.

Divide batter into prepared ramekins. Bake for 10 minutes. Enjoy warm or cold.

YIELD: 2 servings

CHOCOCADO PUDDING

 Fat sub: no oil added, avocado for natural fat

 Corn Free   **Gluten Free**

 Nut Free   **Quick and Easy**

 Soy Free   **Wheat Free**

This guilt-free pudding will have people wondering why all puddings aren't made from avocados! With no added fat or refined sugar, it's okay to get second helpings.

- 12 ounces (340 g) fresh avocado flesh (about 4 avocados)
- 1/4 cup (20 g) unsweetened cocoa powder
- 1/4 cup (84 g) brown rice syrup
- 1/4 cup (84 g) agave nectar to taste
- 2 teaspoons fresh lemon juice
- 1/2 teaspoon pure vanilla extract
  Pinch salt
- 1/4 teaspoon chipotle powder, optional

COMBINE ALL INGREDIENTS in a blender and purée until silky smooth.

Divide among 4 dessert cups and chill before serving.

YIELD: 4 servings

Acknowledgments

Thank you, Amanda Waddell, and Will Kiester at Fair Winds Press for putting up with us once again; to Nancy King for being the best typo slayer; to Rosalind Wanke for her invaluable photography tips; and to Carol Holtz for a beautifully designed book.

The "Besters":
Courtney Blair, Monika Soria Caruso, Michelle Cavigliano, Becca Cleaver, Lisa Coulson, Amy Gedgaudas, Heather Graves, Ricki Heller, Kat Hindmarsh, Jenny Howard, Kathleen Mather, Monique and Michel Narbel-Gimzia, Tami Noyes, JoLynn Ochoa, Kat Raese, Constanze Reichardt, Luciana Rushing, Ashley Stephen, Karen Stewart, Nina Stoma, Liz Wyman, and Andrea Zeichner.

The many businesses and people who supported us when *500 Vegan Recipes* came out:
Josh and Michelle at www.herbivoreclothing.com
Julie and Jay Hasson at www.everydaydish.tv
Lindsey and Melanie at www.shophumanitaire.com
Leigh at www.cosmosveganshoppe.com
Whole Foods Market
Russo's Bookstore

Joni would like to thank Rich, Bryan, Jesus, Matt and the rest of the LAG WFM crew for working around her ridiculous schedule and standing behind her along the way. She would also like to thank her husband, Dan: I love you, baby! And, Celine, there are not even words. . .

Celine would like to thank her husband, Chaz, for sacrificing his girlish figure for the sake of another cookbook; her cats for keeping her insane; her parents for being the most supportive and generous people she's ever known; Afsoon Azizi for being made of kindness; and last but definitely not least, Joni: We did it again!

About the Authors

JONI MARIE NEWMAN, a Southern California native, lives in Orange County with her husband, three dogs, and the cat.

She is the founder of justthefood.com, the author of *Cozy Inside*, and the co-author of *500 Vegan Recipes*. You can get in touch with her at joni@justthefood.com.

CELINE STEEN was born in Switzerland and currently lives in California with her husband and two cats.

She is the founder of havecakewilltravel.com and the co-author of *500 Vegan Recipes*. You can get in touch with her at celine@havecakewilltravel.com.

ENJOY KITCHEN SUCCESS!

THE COMPLETE CHART OF VEGAN FOOD SUBSTITUTIONS

CHAPTER	IF THE ORIGINAL RECIPE CALLS FOR...	REPLACE WITH...
1 **DAIRY**	1 cup (235 ml) buttermilk	Combine 1 tablespoon (15 ml) fresh lemon juice or vinegar, such as apple cider or white balsamic, with 1 cup (235 ml) unsweetened soymilk
	1 cup (235 ml) cow's or goat's milk	1 cup (235 ml) soymilk 1 cup (235 ml) almond milk 1 cup (235 ml) hemp milk 1 cup (235 ml) rice milk 1 cup (235 ml) coconut milk for drinking, such as So Delicious
	1 cup (224 g) dairy butter	1 cup (224 g) non-hydrogenated, nondairy butter, such as Earth Balance 1 cup (224 g) coconut oil ¾ cup (168 g) vegetable shortening
	1 cup (235 ml) heavy cream	1 cup (235 ml) soy creamer 1 cup (235 ml) full-fat unsweetened coconut milk, or coconut cream
	2 scoops ice cream	2 scoops soy ice cream 2 scoops rice ice cream 2 scoops coconut ice cream 2 scoops Vanilla Latte Ice Cream (page 38) 2 scoops Maple Orange Creamy Sorbet (page 38) 2 scoops Pumpkin Ice Cream (page 40)
	1 cup (240 g) sour cream	1 cup (240 g) nondairy sour cream, such as Tofutti or Follow Your Heart 1 cup (240 g) Basic Tofu Sour Cream (page 21)
	1 cup (240 g) yogurt	1 cup (240 g) coconut yogurt 1 cup (240 g) rice yogurt 1 cup (240 g) soy yogurt 1 cup (240 g) Basic Homemade Sorta Yogurt (page 20)
2 **CHEESE**	1 cup (112 g) shredded cheese	1 cup (112 g) nondairy shredded cheese, such as Daiya
	1 slice cheese (for sandwiches)	1 slice Galaxy or Tofutti Cheese Slices (available in several flavors) 1 slice Nutty Pepperjack (page 48)

CHAPTER	IF THE ORIGINAL RECIPE CALLS FOR...	REPLACE WITH...
2 **CHEESE**	1 cup (240 g) cream cheese	1 cup (240 g) nondairy cream cheese, such as Tofutti or Follow Your Heart
	1 cup (235 ml) nacho cheese sauce	1 cup (235 ml) Nacho Queso (page 52)
	1 cup (150 g) feta crumbles	1 cup (150 g) Tofu Feta (page 46)
	1 cup (245 g) ricotta cheese	1 cup (245 g) Tofu Ricotta (page 47)
3 **EGGS**	1 egg in baked goods, for binding (for cookies and cakes)	1½ teaspoons egg replacer powder whisked with 2 tablespoons (30 ml) warm water
		2 tablespoons (16 g) cornstarch or arrowroot, or any starch whisked with 2 tablespoons (30 ml) water
		2½ tablespoons (18 g) flaxseed meal whisked with 3 tablespoons (45 ml) warm water
		¼ cup (60 g) blended silken tofu
		¼ cup (60 g) applesauce, pumpkin, or other fruit or vegetable purée
	1 egg in baked goods, for leavening (for fluffy cakes, muffins, or quick bread)	1 tablespoon (15 ml) mild-flavored vinegar combined with nondairy milk (soymilk yields best results) to curdle and make 1 cup (235 ml)—works best in recipes that involve baking soda
		¼ cup (60 g) nondairy yogurt
		1½ teaspoons egg replacer powder whisked with 2 tablespoons (30 ml) warm water
	1 egg in baked goods, for moisture (for cakes and cookies)	¼ cup (60 ml) coconut milk (for its fat content)
		1 teaspoon oil or nut/seed butter combined with nondairy milks, to make ¼ cup (60 ml)
		¼ cup (60 g) fruit or vegetable purée
	1 egg white	1½ teaspoons egg replacer powder whisked with 2 tablespoons (30 ml) warm water
		Dissolve 1 tablespoon (8 g) agar powder in 1 tablespoon (15 ml) water, whip, chill thoroughly, and whip some more. (We do not recommend this for recipes that call for more than 2 egg whites.)
	1 egg in savory foods, for binding	1½ teaspoons egg replacer powder whisked with 2 tablespoons (30 ml) warm water
		¼ cup (60 g) blended silken tofu
		2½ tablespoons (18 g) flaxseed meal whisked with 3 tablesppons (45 ml) warm water
		2 tablespoons (33 g) tomato paste or other vegetable purée or unsweetened nut/seed butters, as a moisturizing and binding agent, as in Meatfree Balls (page 114)
		2 tablespoons (15 g) flour, starches, bread crumbs, as a binding agent, as in Meatfree Balls (page 114) or Denver Quiche (page 70)
	1 egg in savory foods, for leavening	¼ cup (60 g) blended silken tofu
	Scrambled eggs	Scrambled tofu as in Spinach and Mushroom Tofu Scramble (page 72)
	Hard-boiled eggs	Extra- or super-frim tofu as in Traditional "Egg" Salad (page 68)
4 **MEAT**	8 ounces (227 g) bacon (bits)	8 ounces (227 g) store-bought imitation bacon bits, such as Bacuns
		8 ounces (227 g) Imitation Bacon Bits (page 143)
	8 ounces (227 g) bacon (strips)	8 ounces (227 g) store-bought vegan bacon, such as Lightlife Smart Bacon
		8 ounces (227 g) Black Forest Bacon (page 142)
	1 cup (235 ml) beef broth *(continued on next page)*	1 cup (235 ml) store-bought beef-flavored vegetable broth, such as Better than Bouillon No Beef Base

CHAPTER	IF THE ORIGINAL RECIPE CALLS FOR...	REPLACE WITH...
4 **MEAT**	1 cup (235 ml) beef broth *(continued from previous page)*	1 cup (235 ml) plain vegetable broth 1 tablespoon (15 ml) steak sauce and 1 tablespoon (15 ml) soy sauce mixed with 1 scant cup (205 ml) plain vegetable broth
	8 ounces (227 g) beef (ground)	8 ounces (227 g) store-bought ground beef substitute 8 ounces (227 g) TVP reconstituted with beef-flavored broth
	8 ounces (227 g) beef (steaks or burgers)	8 ounces (227 g) store-bought veggie burgers 8 ounces (227 g) Portobello mushrooms 8 ounces (227 g) Baked Seitan Cutlets (page 130) 8 ounces (227 g) Basic Traditional Boiled Seitan (page 110)
	8 ounces (227 g) beef (strips)	8 ounces (227 g) Basic Traditional Boiled Seitan (page 110) cut into strips 8 ounces (227 g) Portobello mushrooms cut into strips
	8 ounces (227 g) chicken (breasts)	8 ounces (227 g) store-bought products, such as Gardein or Match Foods 8 ounces (227 g) Baked Seitan Cutlets (page 130)
	1 cup (235 ml) chicken broth	1 cup (235 ml) chicken-flavored vegetable broth, or plain vegetable broth
	8 ounces (227 g) chicken (strips or nuggets)	8 ounces (227 g) chicken-style store-bought strips or nuggets, such as Gardein or Morningstar
	8 ounces (227 g) chicken or turkey (ground)	8 ounces (227 g) TVP reconstituted in chicken-flavored broth
	8 ounces (227 g) chorizo	8 ounces (227 g) store-bought soy chorizo, such as Soyrizo 8 ounces (227 g) Seitan Chorizo Crumbles (recipe included in *500 Vegan Recipes*)
	8 ounces (227 g) crab (lump)	8 ounces (227 g) store-bought imitation crab, such as Match Foods
	8 ounces (227 g) deli meats (slices)	8 ounces (227 g) store-bought vegan deli meat slices, such as Tofurky or Yves 8 ounces (227 g) Basic Traditional Boiled Seitan (page 110) thinly sliced
	8 ounces (227 g) fish (fillets)	8 ounces (227 g) Fish-y Sticks (page 145) 8 ounces (227 g) White Bean Cutlets (page 220)
	8 ounces (227 g) hot dogs	8 ounces (227 g) store-bought vegan hot dogs, such as Tofurky or Tofupups 8 ounces (227 g) All-American Hot Dogs (page 136)
	8 ounces (227 g) pepperoni	8 ounces (227 g) store-bought vegan pepperoni, such as Yves 8 ounces (227 g) TVP or Seitan Pepperoni
	8 ounces (227 g) canned tuna	8 ounces (227 g) store-bought mock tuna, such as Tun-o 8 ounces (227 g) Happy Sea Tempeh Salad (page 146)
5 **BY-** **PRODUCTS**	1 cup (224 g) butter or lard	1 cup (224 g) nondairy butter, solid coconut oil, or vegetable shortening
	Food dye	Beet, tomato, spinach, carrot, blueberry juice or powder Store-bought natural food dye, such as India Tree
	Gelatin	Agar, fruit pectin, carrageenan, locust bean gum
	1 cup (235 ml) honey	1 cup (235 ml) agave nectar, barley malt, pure maple syrup, blackstrap or regular molasses
6 **GLUTEN**	1 cup (125 g) all-purpose flour	1 cup (120 g) Homemade All-Purpose Gluten-Free Baking Mix (page 173) or store-bought GF baking mix

CHAPTER	IF THE ORIGINAL RECIPE CALLS FOR...	REPLACE WITH...
6 **GLUTEN**	1 cup (200 g) pearl barley, uncooked	1 cup (185 g) uncooked brown, white, or wild rice
		1 cup (168 g) uncooked quinoa
		1 cup (200 g) uncooked millet
		1 cup (193 g) uncooked amaranth
	1 cup (19 g) puffed wheat	1 cup (16 g) puffed brown or regular rice, quinoa, millet, amaranth
	8 ounces (227 g) seitan	8 ounces (227 g) tofu, tempeh or TVP
7 **SOY**	1 cup (155 g) shelled edamame	1 cup (240 g) lima beans, cannellini beans, garbanzo beans, or fava beans
	1 cup (235 ml) soymilk	1 cup (235 ml) other nondairy milk, such as almond, coconut, rice, oat, or hemp. (Be mindful that no soy is hidden in store-bought nondairy alternatives to soymilk)
	8 ounces (227 g) tofu, tempeh, or TVP	8 ounces (227 g) seitan, beans, or other legumes (except soybeans or edamame), chickpea tofu, peanut tofu
	Soy-based cheese	Rice-based cheese, Daiya, nutritional yeast–based dishes, or nut-based cheeses, such as Nutty Pepperjack (page 48)
8 **SUGAR**	1 cup (225 g) brown sugar, packed	1 cup (192 g) dark evaporated cane juice, such as Sucanat
		1 cup (235 ml) liquid sweetener, decrease other liquid by ⅓ cup (80 ml) or add extra ¼ cup (30 g) flour
	1 cup (235 ml) liquid sweetener	1 cup (235 ml) agave nectar, barley malt, pure maple syrup, blackstrap or regular molasses, brown rice syrup, fruit syrup
		1 cup (192 g) dry sweetener, increase liquid by ⅓ cup (80 ml), or decrease flour by ¼ cup (30g)
		Fruit purée, fruit juice concentrate for nondairy yogurts or smoothies, to taste
	1 cup (192 g) white sugar	1 cup (192 g) evaporated cane juice
		1 cup (235 ml) liquid sweetener, decrease other liquid by ⅓ cup (80 ml) or add extra ¼ cup (30 g) flour
	1 cup (120 g) powdered sugar	½ cup plus 2 tablespoons (120 g) evaporated cane juice ground in a very dry blender until powdery
		1 cup (120 g) store-bought organic powdered sugar
9 **FAT**	½ cup (120 ml) melted dairy or nondairy butter	½ cup (120 ml) grapeseed, olive, peanut, or melted coconut oil (depending on use, choose mild-flavored oils)
	½ cup (112 g) dairy butter	½ cup (112 g) non-hydrogenated, nondairy butter
	½ cup (112 g) dairy butter (in baked goods)	¼ cup (60 g) unsweetened applesauce or other mild-flavored fruit or vegetable purée (dried plums, avocado, beans…) plus ¼ cup (56 g) nondairy butter
	½ cup (112 g) dairy butter (in cakes)	¼ cup plus 3 tablespoons (105 ml) vegetable or other mild-flavored oil
	½ cup (112 g) dairy butter (in cookies)	¼ cup (60 ml) vegetable or other mild-flavored oil, may change the texture of cookies
	½ cup (112 g) shortening	½ cup (112 g) nonhydrogenated vegetable shortening
	½ cup (60 ml) vegetable oil (in cakes)	¼ cup (60 ml) oil plus ¼ cup (60 g) unsweetened applesauce or other mild-flavored fruit or vegetable purée (avocado, beans…)
	Vegetable oil (to sauté vegetables)	Pinch salt to release moisture from veggies, plus vegetable broth, water, wine, or beer as needed
		Non-stick cooking spray
		Fresh herbs and seasonings to boost flavor

The Complete Chart of Vegan Food Substitutions   269

INDEX